D0205444

From the

Forest

to the Sea

*The Ecology of Wood in
Streams, Rivers, Estuaries,
and Oceans*

By

Chris Maser

James R. Sedell

S^t_L

Printed and bound in the U.S.A. Printed on acid-free paper.

Library of Congress Cataloging-in-Publication Data

Maser, Chris.
    From the forest to the sea : the ecology of wood in streams, rivers, estuaries, and oceans / by Chris Maser and James Sedell.
      p. cm.
    Includes bibliographical references and index.
    ISBN 1-884015-17-4
    1. Aquatic ecology.   2. Wood—Environmental aspects.   3. Wood—Biodegradation—Environmental aspects.   I. Sedell, James R.
II. Title.
QH541.5.W3M365   1994
574.5′263—dc20

                        94-11654
                        CIP

Phone: (407) 274-9906
Fax: (407) 274-9927

S$_{\text{L}}^{t}$

Published by
St. Lucie Press
100 E. Linton Blvd., Suite 403B
Delray Beach, FL 33483

DEDICATION

To Robert F. Tarrant

Who over the course of many years has always believed in me even when he could neither see nor understand the vision of my dreams for a whole, healed world

Chris

To my father, John C. Sedell

Whose love of the land and its forests and waters was transferred to me during numerous camping trips while I was growing up

Jim

To Ruth Turner

The preeminent pioneer of wood utilization as a basis of a detrital food chain on the deep-sea floor. Only through the results of Ruth's research could one of the most important parts of this book be written.

CONTENTS

FOREWORD

The forest and the sea have always been connected in human mind and myth, as well as ecologically. For most of our history, we have gone "down to the sea in ships" made from trees, emulating driftwood, which was probably the original model for the whole process. In any event, the knitting together of sea and land by rivers, logs, fish, soil, and tides is basic to the ecology of all coastal margins where forests occur. Because most great rivers rise in forested places and run to the sea, the connection is fundamental and well-nigh universal and extends far inland and upland from the coast.

How interesting, then, and how devastating a commentary on the current state of our disciplinarily fragmented science, that the people who study forests and the people who study salt water rarely interact. How peculiar that it has been only in the last two decades that we realized that much of the "sediment" in the coastal rivers that found its way to estuaries and coastal waters of the Pacific Northwest, the setting for this marvelously inquisitive book, was solid wood.

The authors offer a combined historical and ecological perspective rare in "nature" books. As they recount, their puzzlement at the great decline in the amount of wood on the beaches of the Pacific Northwest during their lifetime made them curious, and this put them on a trail that led them back to earliest European settlement and into contemporary oceanography laboratories. The book is idiosyncratic in the good sense of the word; that is, it is obviously the product of the authors and no one else, which you will quickly realize if you have the pleasure of knowing them.

This is a book everyone should read. It describes a connection that is both vital to ecological and commercial health and one which we have chosen, at least until recently, to know little about. The connection between logs, rivers, and ocean life is both easier to see and easier to understand, once explained, than the esoterica of ants in the Amazon or "biogeochemical cycles." It illustrates deftly that it is the connections that count. We ignore them at our peril. The odds are that you won't again ignore this one if you read on.

John C. Gordon
Pinchot Professor of Forestry
Yale School of Forestry and
Environmental Studies
April 1994

PREFACE

We grew up in Oregon and over the years spent considerable time at the ocean. As boys we remember the huge piles of driftwood along the beaches, piles that seemed to grow with each winter storm. In fact, one of the challenges of even getting to the sandy shores of the Pacific was having to climb over the jumbled mountains of driftwood. There was so much wood, ranging from small branches to boards to whole trees, that we could build shelters from the wind that easily held fifteen or more people. Enormous piles of driftwood were simply part of the beach, a part we took as much for granted in our growing up as clean air and safe sunshine.

Then we were grown, and, suddenly, the driftwood mountains were gone. Where did they go? What happened to them? When did they disappear and why?

To us, the initial puzzle was: How could the mountains of driftwood we so clearly remember have vanished without our noticing? The answer to this question lies in our limited human ability to understand cause and effect and to sense continual change.

It is easy to sense change—the growing light at sunrise, the gathering wind before a thunderstorm, the changing seasons of spring's new leaves, summer's swaying flowers, autumn's ripe harvest, winter's stark, naked trees and chilling winds. Some can perceive longer-term events and remember that there was more snow than usual last winter or that spring came early this year. It is an unusual person who can sense, with any degree of precision, the changes that occur over the decades of his or her life.

Seen only throughout the span of a human lifetime, the world appears static, and we typically underestimate the degree to which change occurs. Slow changes are difficult to detect, especially those that are part of people simply living their daily lives and interacting with their immediate environment. People are even more limited in their abilities to interpret the relationship of cause and effect—those subtle processes acting quietly and unobtrusively over decades that reside hidden in what Professor John Magnuson calls the "invisible present."

Within the time scale of a person's lifetime and the lifetimes of his or her children and grandchildren, entire ecosystems change. Ironically, however, it is within this time scale that people are most blind to changes occurring around them.

At first, we merely felt an acute sense of loss, a sense that as we grew up something that had always seemed a constant in our world was changing unbeknownst to us. That was disquieting in itself. Then we began investigating the other questions: Where did the huge piles of driftwood go? What happened to them? When did they disappear and why?

The more we pondered and teased these questions, the more urgency we felt to press on because we began to understand that what society is doing to the forests of western Oregon and the Pacific Northwest is having devastating effects, not only on our local streams, rivers, estuaries, and beaches but also on the Pacific Ocean itself, effects felt at least as far away as Hawaii.

The forests, which through the millennia have fed the ocean energy in the form of driftwood, especially large driftwood, are rapidly disappearing. Without large driftwood, the ocean—from its shallowest estuary to the deepest part of its floor—is being deprived of terrestrially based carbon, a vital source of energy, and habitat.

As our studies progressed, other questions came to the fore: What does it mean to starve the ocean? What exactly is driftwood? Most people associate driftwood with the ocean itself, but where does it really come from? What effect does logging, especially of old-growth forests, have on driftwood? What does driftwood do? Is it important? Can the growing absence of driftwood be part of the problem of our declining runs of wild salmon? What role does driftwood play in the life cycle of pelagic tuna?

Although the questions are endless, we reached a point when necessity's sharp, insistent pinch to make available our accumulated information became the overriding consideration. It became increasingly clear that there is little time left to begin repairing the damage done, some of which may already be irreversible. Despite our sense of urgency, however, we also recognized that our presentation must be as simple to understand as possible and still be ecologically accurate. This has proven a most difficult task and has taken us five years because of the amount of information and the sheer complexity of the subject. We have done our best to keep our personal biases to ourselves, although we are not sure we always succeeded.

Driftwood, wood carried by water from the forest to the sea, is a critically important source of habitat and food for the marine ecosystem, including the deep-sea floor. Even during its seaward journey, driftwood is both habitat and a source of food for a multitude of plants and animals, both aquatic and terrestrial. In addition, some driftwood controls stream velocities, stabilizes stream banks, makes waterfalls and pools, and creates and protects fish spawning areas. Other driftwood protects the encroachment of vegetation on floodplains and allows forests to expand. In short, driftwood makes a vital contribution to the health of streams, rivers, estuaries, and oceans, not just in the Pacific Northwest but worldwide.

The natural processes by which wood disappears from streams and rivers have positive effects on the ecosystem. Human activities, on the other hand, such as cleaning streams, logging, and firewood cutting, have had negative effects over the last several decades. The consequences of these actions, however, are both little understood and far reaching.

To understand what severing the connection means ecologically and to better grasp our options for the future, we have written *From the Forest to the Sea* in three sections, beginning with the present and working back through the past.

In Part I, The Present: A Severed Connection, some current actions, including damming streams and rivers, which prevents the driftwood they once carried from reaching the sea, are considered. Without driftwood, stream channels and coastal sand dunes are destabilized and wood is eliminated from the food chains of estuaries and the deep-sea floor. Once-plentiful driftwood is rapidly being replaced with such human-derived garbage as non-biodegradable plastics.

In Part II, From the Forest to the Sea: Source, Function, and Transport of Driftwood, the journey taken by water and wood from the mountains to the ocean is told. Through this story, derived from over a decade of research by many people, the vital role of driftwood in streams, rivers, estuaries, and the open ocean is described. Although no specific dates are given, the story is set prior to the settlement of the Pacific Northwest by

Europeans in order to establish a sense of how the presettlement ecosystem might have functioned and to set the stage for Part III.

In Part III, From the Sea to the Forest: Settlement, a look at the collision of European culture and settlement with the environment of the New World is provided. Settlement, which began at the coast and progressed inland up the rivers and streams, began immediately to affect the waterways and ancient forests of the Pacific Northwest, contributing to current environmental crises. Although we look at the Pacific Northwest as a specific chapter in the history of the New World, the principles discussed apply to all of North America—and indeed to all the world.

Chris Maser
Jim Sedell

ACKNOWLEDGMENTS

From the Forest to the Sea has been an interesting, exciting, and difficult book to write, because it has so many individual and interwoven threads to keep track of. The threads begin in geological time as the various physical and biological cycles that create and shape the land, the forest, and the streams. Then, where this book picks up, the threads extend as forest trees enter the streams as driftwood, where they mix with the life cycles of a great variety of animals that use the wood both as habitat and food. From streams, the driftwood journeys down rivers into estuaries, where other forms of life use it as habitat and as food. From the estuaries, some driftwood is carried into the sea, where it floats to the Hawaiian Islands and is made into canoes or sinks to the deep-sea floor as oases of life-giving habitat and food.

Because of the complexity of the story, we had to draw on the research of a wide variety of people, each of whom has added an individual gift of knowledge to our understanding of driftwood's journey by publishing her or his studies in scientific literature. In addition to information from scientific literature and our own scientific studies, the following people graciously read the manuscript for clarity and accuracy and made innumerable improvements: Louis and Muriel Barron (Downsville, New York), Will Moir (USDA Forest Service, Rocky Mountain Research Station, Fort Collins, Colorado), Jean Holt (Science Editor, USDI National Park Service, Corvallis, Oregon), and Barbara Bash (Kingston, New York).

Special thanks are extended to Robert F. Tarrant (Former Director of the USDA Forest Service Pacific Northwest Research Station, Portland, Oregon, and Professor of Forest Science, Department of Forest Science, Oregon State University, Corvallis), who read through the entire manuscript twice at vastly different stages in its preparation.

Thanks are offered to Jefferson J. Gonor (Department of Oceanography, Oregon State University) and Patricia Benner (Oregon State University), who were co-authors with us on the initial scientific report *From the Forest to the Sea, A Story of Fallen Trees*, on which some of the marine portion of the present story is based.

Thanks are also extended to the following people:

Norman Anderson (Oregon State University), a friend and colleague, openly provided both information and photographs of the natural history of aquatic insects and their interactions with wood in streams.

Patricia Benner (Oregon State University) graciously provided both a previously unpublished figure of the Willamette River and her historical research that reconstructs the profile of driftwood in estuaries and on beaches of the Pacific Northwest.

The Stream Team (Oregon State University), a generous and creative group lead by Stan Gregory, which over the past two decades has defined the relationship of the stream to its forest at the H. J. Andrews Experimental Forest in western Oregon.

Jim Leadon, Publications Editor for Agricultural Communications at Oregon State University, provided the original art on which the figures of the gribbles and shipworms are based.

Lewis Nelson, owner of Scott Photo, Corvallis, Oregon, provided both his own photographs and took great pains in crafting the photographs used in this book.

Dan Pool lent his artistic talents to creating many of the figures in this book.

Fred Swanson (Pacific Northwest Research Station of the U.S. Forest Service) generously provided photographs and ideas about landslides, earthflows, and the role of wood in shaping the geomorphology of streams.

Ruth Turner (Harvard University) not only did the seminal work on deep-sea wood-borers and wood islands but also was a generous and enthusiastic source of photographs and ideas about the ecological role of wood in the deep-sea environment.

And finally, Zane Maser and Ellen Sedell, our intrepid wives, have been strong supporters of our efforts and have been helpful and patient in many unseen ways as we worked on this book.

The forests are dying, the rivers are dying,
and we are called to act.
To return Earth to harmony is to restore
the harmonious principles within ourselves
and to act as responsible caretakers
to save the forests and the waters for future generations.

Dhyani Ywahoo

The care of rivers is not a question of rivers,
but of the human heart.

Tanaka Shozo

PART I

THE PRESENT: A SEVERED CONNECTION

CHAPTER 1

EUROPEAN SETTLEMENT AND LAISSEZ-FAIRE CAPITALISM

European settlement along the east coast gave birth to greed that drove men across the vast North American continent in a frenzy of conquest and exploitation. The Constitutional Convention of 1787 prepared the way for westward expansion across what is now the United States. By 1812, the overriding emphasis of human endeavors in the Pacific Northwest was focused on the battle for Nature's bounty. It was in the 1840s, in what are now Oregon and Washington, that human manipulation of the environment began to change the forest in ways never before seen. Finally, World War II set the technological stage for the systematic alteration, which often constitutes destruction, of the forests, streams, rivers, estuaries, and oceans faster and more completely than at any time in history.

Logging in what is now the Willamette National Forest in western Oregon was first recorded in 1875. During the first three decades of this century, 90 percent of the timber cut was still readily accessible near rivers and streams below a 4000-foot elevation. By the 1970s, 65 percent of the timber cut occurred above a 4000-foot elevation, and the trees felled became progressively younger and smaller. Today, to maintain the same volume of wood fiber cut below 4000 feet in elevation, five times the number of acres above the 4000-foot level is being cut.[1]

In just 142 years, from the start of Lewis' and Clark's epic travels in 1803 to the end of World War II in 1945, we in the United States have achieved the technological capability of disarranging and disarticulating the basic biological functioning of the world. Today, less than 50 years since the end of World War II, the demands of human society are causing rapid, global deforestation. In the process, forests are both physically and ecologically uncoupled from streams, rivers, estuaries, and oceans of the world.

Today, for example, driftwood is prevented from even beginning its journey by the removal of as much wood as possible from the forests to be used as a product for human consumption, lest it remain as an "economic waste" (Figures 1.1 and 1.2). The little driftwood that even begins its journey is prevented from completing it by dammed rivers

FIGURE 1.1
Clearcut logging prevents trees from becoming driftwood. (USDA Forest Service photograph by Tom Spies.)

FIGURE 1.2
Today, small streams (left center) and large streams (bottom left) have only little patches of forest feeding them driftwood. (Photograph by Arthur McKee.)

(Figures 1.3 and 1.4). Thus is severed the connection between forest and sea.

In addition, what now substitutes for driftwood in the ocean and on beaches is such nonwooden human garbage as metal, glass, rubber, plastic, oil, bilge, chemical effluents, medical and household wastes, and raw human sewage. In 1987, for example, 17 tons of human garbage were cleaned from Oregon beaches. (In 1989, 26 tons of garbage were collected, 17.5 tons in 1991, and 36 tons in 1994). Six tons of human garbage were collected from Washington beaches, 75 tons from California beaches, 306.5 in Texas, 36.8 in Hawaii, 200 in Louisiana, and 40 in New Jersey. This garbage comes from such sources as recreation and commercial boats; commercial, military, and research ships; beach-goers; offshore oil and gas rigs; shore-based solid wastes; manufacturing; and sewage treatment plants.

The plastics dumped into the marine environment cannot replace

FIGURE 1.3
Driftwood collected behind a log boom to keep it from coming in contact with a dam, effectively preventing it from ever reaching the sea. (Photograph by Chris Maser.)

driftwood. Research involving deep-sea wood-borers and *in situ* experiments with "wood islands," bags of algae, eelgrass, and dead fish, as well as research describing life around the thermal vents, suggest that the most important factor supporting the evolution and maintenance of diverse animal communities in the deep sea is a varied and continued source of food, including driftwood.

Off the coast of North America, the supply of driftwood for food on the ocean bottom is both dwindling and becoming more erratic. For the first time in the evolutionary history of deep-sea animals, the availability of food has become unpredictable. If the coastal mangrove forests continue to be destroyed through deforestation, the last direct link between the forest and sea will be severed. Then the deep-sea wood-dependent species of the world will shrink in numbers, the areas they inhabit will decrease, and some species will become extinct. This partial or total

FIGURE 1.4
Driftwood hauled out of the containment boom and piled to be cut as firewood.
(Photograph by Chris Maser.)

extinction of species would have an immeasurable effect on the world's oceans.

Both these individual species and the ecological link between forest and sea are currently in jeopardy. The elimination of driftwood is a significant factor in the severing of this vital connection. To really understand the severity of the disconnect, however, it is necessary to examine the ecological role of wood in streams, rivers, estuaries, and oceans.

PART II

FROM THE FOREST TO THE SEA: SOURCE, FUNCTION, AND TRANSPORT OF DRIFTWOOD

CHAPTER 2

THE STREAMS

THE STREAM-ORDER CONTINUUM

A first-order stream is the smallest undivided waterway or headwaters. Where two first-order streams join, they enlarge into a second-order stream. Where two second-order streams come together, they enlarge into a third-order stream, and so on.

The concept of stream order is based on the size of a stream in terms of the cumulative volume of water, not just on the given orders of the streams that converge. For example, a first-order stream can join either with another first-order stream to form a second-order stream or it can enter directly into a second-, third-, fourth-, fifth-, or even larger-order stream. The same is true of streams of all orders.

Streams are the arterial system of the land (Figure 2.1). They form a continuum of physical environments and associated aquatic and terrestrial plant and animal communities. This continuum is a longitudinally connected part of the ecosystem in which downstream processes are linked to upstream processes (Figure 2.2).

The stream continuum begins with the smallest stream and ends at the ocean. It includes the available food resources for the animals inhabiting the continuum, which range from invertebrates to fish, birds, and mammals. More important, however, is the role played by streamside vegetation in the control of water temperature, stabilization of stream banks, and production of food. Streamside vegetation is also the primary

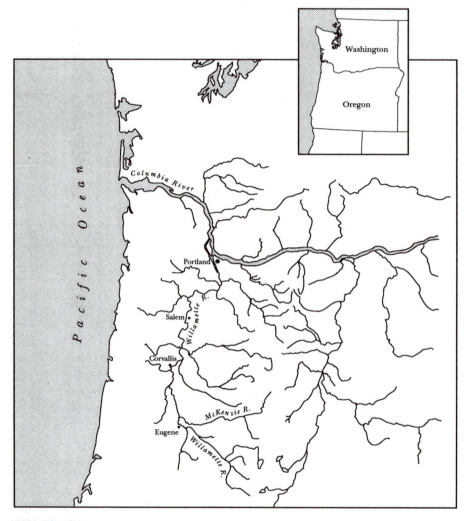

FIGURE. 2.1
Map showing origin and route of streams carrying driftwood that are mentioned in the text.

source of large organic debris (primarily driftwood), such as tree stems at least eight inches in diameter with their rootwads attached or tree branches greater than eight inches in diameter.

Forests adjacent to streams supply such driftwood as stems (Figure 2.3), rootwads (Figure 2.4), and large branches. Erosion also con-

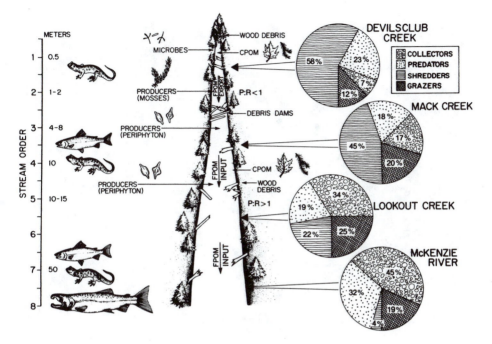

FIGURE. 2.2
River drainage system characterized as a continuum from small streams to large rivers. First- to eighth-order streams display width, dominant predators, groups of producers, production:respiration (P/R) ratios, importance of driftwood, and changing proportions of invertebrate functional groups. CPOM = coarse particulate organic matter, FPOM = fine particulate organic matter. (Adapted from Triska, F.J., J.R. Sedell, and S.V. Gregory. 1982. Coniferous forest streams. in *Analysis of Coniferous Forest Ecosystems in the Western United States,* Edmonds, R.L. (Ed.), US/IBP Synth. Ser., Hutchinson Ross Publ. Co., Stroudsburg, Pa., pp. 292–332.)

tributes driftwood and other debris to the stream and accounts for the water transport of the individual pieces of debris. Driftwood in streams increases the diversity of habitats by forming dams and their attendant pools and by protecting backwater areas. Driftwood also provides nutrients and a variety of foundations for biological activity, and it both dissipates the energy of the water and traps its sediments.

Driftwood floats downhill from the forest to the sea; as the river gets larger, the driftwood gets smaller. First-, second-, and third-order streams feed small rivers, such as the McKenzie River, in western Oregon

FIGURE 2.3
Stem of an old-growth Douglas-fir. (Photograph by Chris Maser.)

(see Figure 2.1), with partially processed food, the amount of which becomes progressively smaller as it travels down the continuum of the river system.

The food, which provides energy to stream organisms in the ancient forest, is mostly litter, such as leaves, needles, cones, twigs, bark, and wood. The large stems of the ancient trees require special consideration because they enter streams infrequently rather than annually. Being somewhat sedentary, they physically shape small streams. In addition, large wood is biologically processed and broken down in place, unless it is flushed downstream in a rare debris torrent (Figure 2.5).

The input of organic debris into streams and its export from streams is predictable. Only 18 to 35 percent of the organic debris that falls or slides into small streams each year is flushed downstream. Streams in the ancient forest retain most of the debris. Some 60 to 70 percent of the annual input of organic debris remains long enough to be biologically useful to the organisms living in the streams. Large dams of woody debris act as sieves and zones of deposition for fine organic debris; such

FIGURE 2.4
Rootwad of an old-growth Douglas-fir. (Photograph by Chris Maser.)

dams also allow time for microbes to colonize the debris and for insects to process it.

Processing of organic debris includes digestion by bacteria, fungi, and insects and physical abrasion against such things as the stream bottom and its boulders. In all cases, organic debris is continually broken into smaller pieces, which makes the particles increasingly susceptible to microbial consumption.

The amount of different kinds of organic debris processed in a reach of stream (a stretch of water that is geomorphically defined, such as between two bends in a stream, river, or channel or between the start of a meadow or canyon) depends on the quality and the quantity of nutrients in the debris and on the stream's capacity to hold fine particles long enough to complete their processing. The debris may be fully utilized by the biotic community within a reach of stream or it may be exported downstream.

Debris moves fastest through the system during high water and is not thoroughly processed at any one spot. The same is true in streams that do

FIGURE 2.5
Track of debris torrent on a small stream. (USDA Forest Service photograph by Fred
J. Swanson.)

not have a sufficient number of in-stream obstacles to slow the water and
act as areas of deposition, sieving the incompletely processed organic
debris out of the current. Small streams thus feed larger streams and
larger streams feed still larger ones.

In small streams flowing through the ancient forest, a large proportion of the basic food resource for invertebrates is derived from leaves
and wood. As the stream gets larger, the influence of the forest diminishes and the stream's source of energy is derived more from aquatic
algae and less from organic debris of the terrestrial forest. The greatest
forest influence is in first-order streams, but the greatest diversity of both
debris inputs and habitats is found in third- to fifth-order streams and
large rivers with floodplains.

Within the ancient forest, wood constitutes between 50 and 70 percent of the total organic debris available to microbes and invertebrates in
small streams, including very fine particles derived almost exclusively
from the massive stems of the ancient trees themselves, but this does not

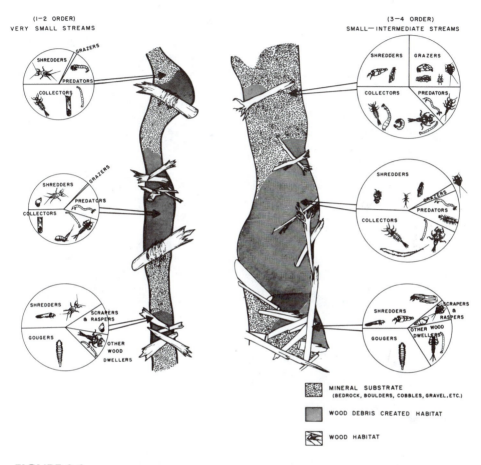

(1–2 ORDER)
VERY SMALL STREAMS

(3–4 ORDER)
SMALL–INTERMEDIATE STREAMS

MINERAL SUBSTRATE
(BEDROCK, BOULDERS, COBBLES, GRAVEL, ETC.)

WOOD DEBRIS CREATED HABITAT

WOOD HABITAT

FIGURE 2.6
Small and intermediate streams showing contrasting proportions of invertebrate functional groups associated with driftwood. (Adapted from Anderson, N.H. and J.R. Sedell. 1979. *Annu. Rev. Entomol.* 24:351–377.)

include the sound portions of the stems. It is not surprising, therefore, that invertebrates in the smallest streams flowing through the ancient forest have evolved to gouge, shred, and scrape wood and leaves and to gather fine organic particles (Figure 2.6).

Small, first-order, headwater streams largely determine the type and quality of the downstream habitat. These and second-order streams are influenced not only by the configuration of surrounding land forms but also the live and dead vegetation along their channels. The vegetation of

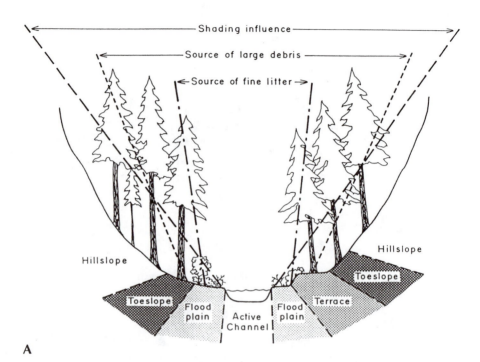

A

RIPARIAN VEGETATION

SITE	COMPONENT	FUNCTION
above ground- above channel	canopy & stems	1. Shade-controls temperature & in stream primary production 2. Source of large and fine plant detritus 3. Wildlife habitat
in channel	large debris derived from riparian veg	1. Control routing of water and sediment 2. Shape habitat-pools, riffles, cover 3. Substrate for biological activity
streambanks	roots	1. Increase bank stability 2 Create overhanging banks-cover 3. Nutrient uptake from ground and stream water
floodplain	stems & low lying canopy	1. Retard movement of sediment, water and floated organic debris in flood flows

B

FIGURE 2.7
Riparian vegetation variously interacts with a stream. (A) Trees outside the riparian floodplain may have a direct influence on the stream. (B) Different components perform different functions.

the streamside forest is called riparian vegetation and interacts in many ways with the stream (Figure 2.7A and B).

Dead vegetation (primarily parts of or whole fallen trees, collectively termed large woody debris) is common in streams flowing through the now-rare pristine forest and often covers up to 50 percent of a stream's channel. The canopy of vegetation, when undisturbed, shades the streamside. The energy of the flowing water is continually dissipated by driftwood in the channel, slowing erosion and fostering the deposition of inorganic and organic debris.

Because these small streams arise in tiny drainages with a limited capacity to store water, their flow may be scanty or intermittent during late summer and autumn. During periods of high flows in winter and spring, however, they can move large amounts of sediment and driftwood.

First- and second-order streams are loaded with driftwood and have many wood-gouging beetle larvae, leaf-shredding stoneflies (Figure 2.8), and snails. The small particles of organic debris sink when trapped by

FIGURE 2.8
Adult stonefly. (Photograph courtesy of Norman H. Anderson.)

large wood and are gathered and eaten by mayflies and midges as well as a copepod, which in this case is a bottom-dwelling invertebrate.

The green plants, the primary producers in streams, also vary widely, as determined by the size of stream. In first- and second-order streams, moss cover is generally greater than 20 percent of the stream area and is located primarily on wood, bedrock, and boulders. These small streams have a sparse diatomaceous flora and a patchy blue-green algal community, which is intimately associated with the mosses.

Third- and fourth-order streams are generally richer than smaller streams in types and numbers of organisms and have a greater combined weight of living tissue or biomass. This diversity and biomass includes a rich variety of insects and populations of vertebrates, such as cutthroat trout, tailed frogs, and Pacific giant salamanders (Figure 2.9), which is due in part to the increased importance of algae as a source of energy.

The algal community is well developed and widely spread throughout these larger streams. The moss community, on the other hand, generally occupies little of the stream area, being confined mostly to wood, bedrock, and large boulders.

Large driftwood creates two types of habitat within each stream: the wood itself and wood-created environments, such as pools where organic debris is deposited (Figure 2.10). Relative proportions of wood-related and other habitats vary markedly with the size of streams. In the smallest streams, 50 percent or more of the area may be occupied by

FIGURE 2.9
Pacific giant salamander. (Photograph by Robert M. Storm.)

FIGURE 2.10
Large driftwood in stream forming two habitats: the wood itself and the associated pool, riffle, and water–stream bank interface. (Photograph by Chris Maser.)

wood and wood-related habitats, as compared with 25 percent in third- and fourth-order streams.

A striking characteristic of first- and second-order streams flowing through the ancient forest is the amount of dead wood in them, ranging from twigs to ancient Douglas-fir trees themselves. The processes through which large wood is transferred from forests into the stream channels vary considerably with such factors as species of trees growing alongside a stream, soil stability, shape of the valley, and type of climate. Large wood is transferred into stream channels via several processes, including caving of stream banks, uprooting of trees by wind, collapse of trees under snow or ice, snow avalanches, and mass soil movements (Figure 2.11). Each of these processes contributes driftwood in a different manner to the channel of a stream. These processes can generally be placed within two groups, each of which depends on the final transfer mechanism.

FIGURE 2.11
Western redcedar being split apart because its roots are half in stable soil and half in an unstable earthflow moving slowly downhill. (USDA Forest Service photograph by Fred J. Swanson.)

WHERE DOES DRIFTWOOD ORIGINATE?

The first category of mechanism that transfers large wood from the forest into a stream's channel consists of frequent, chronic inputs of wood, for example, through death of individual trees from disease and insects, occasional uprooting of trees by wind, and the gradual undercutting of a tree's root system by the stream. Contribution of driftwood larger than 20 inches in diameter is common in mature and ancient forests because of whole-tree mortality.

The second category includes intermittent, sizable inputs of wood via large-scale insect outbreaks or widespread disease, extensive areas of wind-thrown trees, debris avalanches (landslides) (Figures 2.12 and 2.13), and massive stream bank erosion during major floods.

FIGURE 2.12
A recent landslide opens the canopy by sweeping away streamside trees. (USDA Forest Service photograph by Fred J. Swanson.)

FIGURE 2.13
A driftwood dam forms, trapping water and sediments upstream. (USDA Forest Service photograph by Fred J. Swanson.)

FIGURE 2.14
Beaver. (Photograph courtesy of the Oregon Department of Fish and Wildlife.)

The role that debris avalanches and torrents play in bringing driftwood from intermittent streams into large streams and rivers has been greatly underestimated. In mountainous, old-growth watersheds, the location of the majority of driftwood aggregations in intermediate-size streams is at the junction of small tributaries where it has entered the intermediate-size stream via a debris torrent. Thus, much of the driftwood in Pacific Northwest streams and rivers has its origin near ridge tops and not immediately adjacent to a permanently flowing stream.

Beavers (Figure 2.14) are major contributors of wood to streams in young forests and to streams with low gradients. Beavers fell trees, gnaw small sticks, or cut branches along streamsides for food (Figure 2.15) and for dam-building materials (Figure 2.16). One beaver may cause several tons of wood to be added to coastal streams annually, manyfold the yearly addition to a stream from litterfall.

Most twigs and branches end up in streams as a result of wind storms that race wildly through the canopy, whipping trees back and forth.

FIGURE 2.15
Wood cut by beaver as food and left in a stream as refuse becomes driftwood. (Photograph by Chris Maser.)

FIGURE 2.16
Beaver's dams put vast amounts of wood in concentrated areas. (Photograph by Chris Maser.)

Occasional storms are strong enough to uproot old trees, with some falling across streams or into them at various angles. Along the edge of the valley floor where smaller ridges are gently rounded, a strong wind, with gusts exceeding 75 miles per hour, can blow over many trees in localized areas and even hide first-, second-, or third-order streams under the trees' fallen stems.

THE FUNCTION OF DRIFTWOOD

The biological and physical functions of driftwood are affected by its interaction with its environment via small-scale through large-scale processes. Small-scale processes affect driftwood by controlling how it decomposes. In water, wood decomposes more slowly than on land because water-logging prevents oxygen from penetrating deeply into the wood

FIGURE 2.17
Some of these fallen trees will become water-logged; others will be submerged only during high water. Both circumstances affect the rate of decomposition. (USDA Forest Service photograph by James R. Sedell.)

(Figure 2.17). This is an important distinction because the wood-decomposing fungi and invertebrates need oxygen to function.

Water-logged parts of large driftwood decompose in thin layers, starting at the wood's outer surface. As the decomposed surface is scraped off or worn away, oxygen penetrates further into the wood, which becomes food for the decomposer organisms.

Concentration of the elements forming essential nutrients increases as large driftwood decomposes. Nitrogen, for example, increases primarily through the biological use of carbon and through nitrogen fixation by microorganisms.

Medium-scale processes deal with the role of driftwood as animal habitat. Animals that associate with driftwood in streams range from those that are restricted by necessity to living on the wood to those that use the wood only opportunistically.

The sequence of colonists on driftwood reflects the wood's stage of decay. New driftwood entering a stream is used primarily as habitat by algae and microbes that in turn provide food for a group of insects functionally called grazers or collectors. This type of feeding softens driftwood enough that it can be abraded and ingested by invertebrates that scrape their food off surfaces.

More importantly, the driftwood becomes habitable for obligate wood-grazing invertebrates (those that are biologically dependent on wood). These include caddisflies and more generalized wood-shredding invertebrates, such as stoneflies, which eat wood infested with fungi. These activities form a sculpted surface texture, which in turn provides habitat for many additional organisms.

A small stream may receive driftwood partially decayed by fungi and other terrestrial organisms. Such preconditioning allows rapid internal colonization by aquatic microbes and invertebrates, speeding decomposition. Driftwood decomposition is more rapid in larger streams during high water because physical abrasion removes softened tissue as the wood is washed about.

Quality, texture, and species of driftwood determine what kinds of organisms can and will colonize it. The degree to which driftwood is water-logged and its stage of decay are also important.

Smooth wood surfaces are suitable for insect attachment and for grazing the film of microbes; soft wood is more easily penetrated by boring organisms and contains fungal mycelia as nutrients. Aquatic invertebrates are therefore functionally classified by how they use the wood: (1) boring or tunneling in wood; (2) ingesting wood by grazing, scraping, or rasping; (3) grazing algal communities growing on wood; (4) attaching to wood or hiding in its grooves; and (5) preying on other wood-using organisms. Classification by functional groupings is interrelated with wood texture, which partly explains why a higher incidence of facultative organisms (those that are opportunistic in their use of wood) is associated with firm, smooth wood and why a higher incidence of obligate organisms is associated with soft, sculptured wood (Figure 2.18).

Many invertebrates associated with the wood's surface are more opportunistic in their selection of feeding sites and habitat than are those associated with the wood's internal areas. Because the majority of invertebrates are housed in the many grooves and crevices, as well as under loose bark on well-conditioned wood, it is likely that these surface structures are more important as protection from the stream's current, suspended sediments, and predation than as sites for obtaining food.

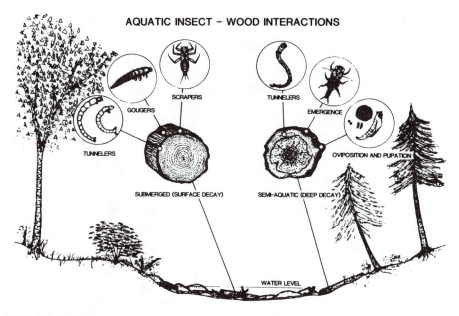

FIGURE 2.18
Interactions between driftwood and aquatic insects in a small stream. (Redrawn from Anderson, N.H. et al. 1984. *Verh. Internatl. Verein. Limnol.* 22:1847–1852.)

In essence, feeding activities of borers, shredders, grazers, and scrapers result in continual decomposition of driftwood. Grazers and shredders (beetles, caddisflies, and some stoneflies) exploit the fungal-enriched area, which contains about five times as much nitrogen as areas of nonenriched wood. Scrapers (mayflies, Figure 2.19) and raspers (snails) also ingest the soft layer as well as the assemblage of minute organisms attached to the surfaces of the submerged wood.

Many species associated with thoroughly decomposed wood are detritus feeders that just happen to be in the wood when it falls into streams. The feeding and burrowing of these organisms reduces the particle size of the woody material and causes it to mineralize.

Net-spinning caddisflies frequently use the surfaces of textured driftwood to attach their nets and to hide (Figure 2.20). Driftwood also directs the stream's flow in a manner that provides net-spinning caddisfly larvae with ideal sites, in terms of both surface structure and water velocity, for attaching their nets from which to filter food from the water.

Numerous species of insects use partially submerged driftwood to

FIGURE 2.19
Adult mayfly. (Photograph courtesy of Norman H. Anderson.)

crawl out of the water so they can emerge as terrestrial adults. Stoneflies and some species of mayflies and dragonflies (Figure 2.21) are among those that prefer wood over mineral surfaces to stage their emergence as terrestrial adults.

The ecological role of wood-eating, aquatic invertebrates is limited, compared with the role of their terrestrial and marine counterparts. In freshwater streams, the creation of feeding grooves in a stick of wood by aquatic beetles, such as the riffle beetle, may be the single most important ecological event, because a stick with a complex surface is likely to

FIGURE 2.20
Nets spun by larval caddisflies provide homes and means of filtering food from water. (Photograph courtesy of Norman H. Anderson.)

FIGURE 2.21
Adult dragonfly. (Photograph courtesy of Norman H. Anderson.)

FIGURE 2.22
Steelhead, a subspecies of rainbow trout, just over one year old. Note spots on dorsal fin. The silvery bubbles on the fish's side are parasites. (USDI Bureau of Land Management photograph by John Anderson.)

support a more diverse plant and animal community than is one with a smooth surface.

Finally, it is necessary to consider large-scale processes in which driftwood both controls the water's flow in a stream's channel and creates and maintains trout and salmon egg-lay habitats and some of the best habitat for survival of young. To live in flowing water, free-swimming species like trout (Figure 2.22) and salmon require sites where food is plentiful for young fish within or close to hiding cover and where little effort is needed to hold a position against the current while feeding.

Those stream riffles that are composed of small gravel have few such sites. Those which do exist are normally occupied by juvenile steelhead trout. Some species, notably coho salmon (Figure 2.23), avoid riffles, relying instead on pools with ample cover provided by large driftwood, where they eat fallen canopy insects or aquatic insects that drift into the pool.

Many fish inhabit driftwood-created pools in streams where the current is slower and the water is deeper than in riffles. Deep pools offer fish a better chance of escaping terrestrial predators and also allow both coexisting species of fish and/or fish of the same species but of different ages to live in layers. Pools created by driftwood dams are often characterized by deep, slow-moving water, low light intensity, and structurally

FIGURE 2.23
Chinook salmon below and coho salmon above by stick. Note the white margins on the coho's fins. Both fish are less than one year old. (USDI Bureau of Land Management photograph by John Anderson.)

complex cover afforded by root masses, deeply undercut banks, and large sunken driftwood. They have the highest use by juvenile coho near the water's surface, cutthroat trout along the bottom, and steelhead trout more than a year old at the upstream head of the pool.

The greater the amount of driftwood in a stream, the greater the number of pools. Around 80 percent of the pools in some small streams are created by driftwood more than ten inches in diameter (Figure 2.24). Driftwood creates a large number of pools in streams of the Pacific Northwest, particularly in first- through third-order streams, where single pieces of large driftwood or accumulations of smaller pieces anchored by a large piece often create a stair-stepped, longitudinal profile of the stream. Such a profile consists of a number of small debris dams with an upstream area that allows the deposit of sediments, the dam itself, and a downstream waterfall ending in a plunge pool scoured out by the force of the falling water (Figure 2.25).

FIGURE 2.24
Driftwood-created plunge pool. (Photograph courtesy of the USDA Forest Service.)

The pools are inhabited by such insects as mayflies, whose nymphal stage feeds on algae, leaves that fall into the water and sink, and fungal mycelia. The fungal mycelia, which can be a high proportion of the diet in some species, are probably gleaned from the surfaces of the rotting leaves and driftwood. Also in the pool are immature dragonflies, stoneflies, caddisflies, and wood-eating craneflies (Figures 2.26 and 2.27). These insects partition the food resource in four different ways. Some shred, some graze, some collect, and some prey on the others.

The shredders depend on large pieces of organic debris, such as leaves, needles, driftwood, and other parts of plants, which they find along the stream edge (Figures 2.28 to 2.32). Grazers remove attached algae, especially when it grows on the surfaces of rocks and fallen trees in the current. Collectors use minute particles of organic debris, generally less than one-sixteenth of an inch in size. They glean this food either by filtering it from passing water or by gathering chosen materials from the stream's bottom. Predators, on the other hand, are adapted through behavior and specialized body parts for capturing prey.

FIGURE 2.25
The stable piece of driftwood forms an obstruction that dissipates some of the stream's energy and creates a pool used by salmonids. (Photograph courtesy of the USDA Forest Service.)

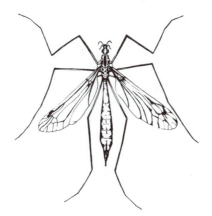

FIGURE 2.26
Adult wood-eating cranefly. (USDA Forest Service illustration by Paula Reid.)

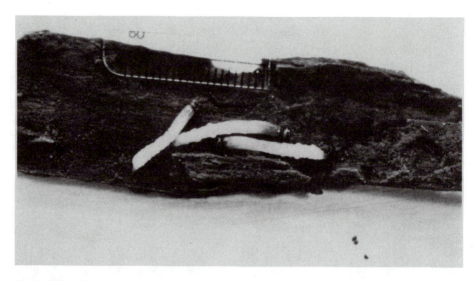

FIGURE 2.27
Larval wood-eating craneflies burrow in decomposing driftwood. (Photograph by Tom Dudley.)

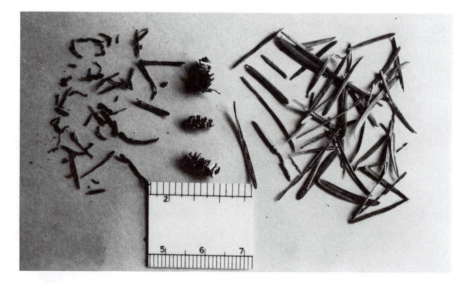

FIGURE 2.28
Larval caddisflies make homes of conifer needles and shredded leaves. (Photograph by Margaret J. Anderson, courtesy of Norman H. Anderson.)

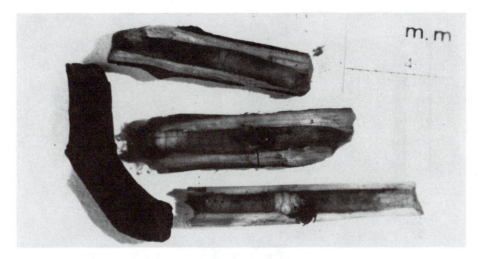

FIGURE 2.29
Larval caddisflies bore holes in small twigs for homes. (Photograph courtesy of Norman H. Anderson.)

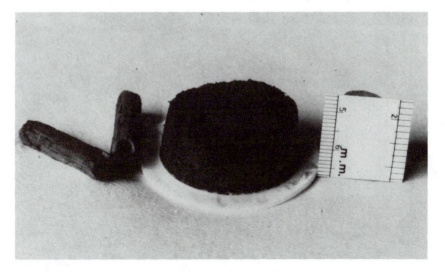

FIGURE 2.30
Larval caddisfly eats shredded leaves and creates a fecal pile like this within days. (Photograph courtesy of Norman H. Anderson.)

FIGURE 2.31
Larval caddisfly's home of conifer needles. (Photograph by J.C. Hodges, Jr.)

FIGURE 2.32
Head of a larval caddisfly. (Photograph by J.C. Hodges, Jr.)

The hairy caddisfly, whose head resembles that of Friar Tuck, is an example of an insect that lives along the stream's margin. Larvae construct their cases from fragments of woody debris like a miniature log cabin that can become buoyant. Thus, when dislodged from the stream's bank into the water, the larvae withdraw into their cases, become quiescent, and float downstream on the water's surface. When they become lodged against the margin of the stream or in the debris floating behind a dam, they crawl out of the water onto leaves or driftwood. In this way, their cases provide a mechanism for dispersal into new habitats by increasing the probability that larvae, displaced into the current of first-order streams by high water from winter storms, will come to rest against suitable habitats along the banks of third-order streams.

Between pools are found several species of riffle beetles, which, as the name implies, live in riffles, generally on stones and such organic debris as driftwood. Somewhat cylindrical, they have very smooth wing covers and are usually less than one-quarter of an inch in length.

One species of riffle beetle is a wood-gouging, wood-eating beetle occurring on submerged or partially submerged driftwood. Adults feed by sweeping the surface of moist driftwood with their mouthparts, which are used for scraping wood rather than gouging it. They eat diatoms, fungal spores, and hyphae, but little woody tissue itself. Adults live only two to three months after a larval life of four to six or more years.

The larvae (Figure 2.33) slice off microscopic pieces of decaying driftwood, but neither produce the enzyme necessary to digest it nor have symbiotic intestinal flora to do the job for them. Instead, they probably obtain nutrition by absorbing substances released by fungi and by digesting and absorbing the contents of fungal and bacterial cells.

The most obvious results of larval feeding are changes in driftwood's surface texture and conversion of the wood into fecal particles. Larval feeding also exposes new, unconditioned driftwood to consumption by wood-decaying fungi, speeding decomposition. The creation of grooves during larval feeding is important because driftwood with a varied and complex surface structure supports a more diverse invertebrate community than driftwood with a smooth surface. Driftwood with gnarled surfaces attracts the water ouzel or dipper (Figure 2.34), which walks under water along the stream's bottom feeding on such aquatic insects as caddisflies and occasionally larval riffle beetles.

Large, stable pieces of driftwood are an important habitat for cutthroat and steelhead trout and coho salmon in winter because they en-

FIGURE 2.33
Larval riffle beetle. (Photograph courtesy of R.W. Henderson.)

FIGURE 2.34
Water ouzel or dipper. (Photograph courtesy of the Oregon Department of Fish and Wildlife.)

hance the use of different habitats within a pool. All species prefer pools during the normal winter flow, but the preference level is determined by the quality of a pool, which in turn is often determined by the abundance of driftwood—the more driftwood, the more fish use a pool. Large, stable pieces of driftwood also tend to attract fish to pools along the edge of a stream rather than to those in mid-channel.

During winter floods, the pool-riffle sequence of a stream's stair-stepped profile becomes a continuous, high-velocity torrent in which there is often little protection for trout and salmon from the moving sediments or swift, turbulent waters. During these brief floods, quiet water refuges are provided almost exclusively by anchored driftwood and standing vegetation in a stream's floodplain (Figures 2.35 and 2.36).

Large pieces of driftwood may also expand the feeding space of a fish by creating or enhancing slow-moving or standing water in organic-rich floodplains, even along the smallest of streams. Almost all juvenile coho salmon and cutthroat trout living in large rivers grow up along the water's edge, often in a wood-rich environment where whole trees or accumulations of driftwood offer protective cover and diversified habitats.

Driftwood in streams generally increases the complexity of habitats by physically obstructing the flow of water. Fallen trees extending partially across a channel deflect the current sideways, which causes it to widen the streambed (Figure 2.37). The sediment deposited and stored adjacent to woody debris in a stream's channel also adds to the hydraulic complexity of the stream; this is especially noticeable in organically rich, wide, shallow channels of low-gradient streams, which have highly diversified riffles and pools. Even when a stream's main channel becomes too wide to permit a tree to span its entirety, driftwood that accumulates along its banks causes meanders to be cut off, often creating well-developed systems of secondary channels.

Driftwood also creates variations in channel depth by causing pools to be scoured downstream from the obstruction to the flow of water. Driftwood therefore creates and maintains a physically diverse habitat by (1) anchoring the position of pools along a channel, (2) creating backwaters along a stream's margin, (3) forming systems of secondary channels in valley floors filled with sediments deposited by their own streams, and (4) varying channel depth.

Thus, fallen trees, which have become driftwood, temporarily create new stream habitats. Eventually, however, a storm comes along with

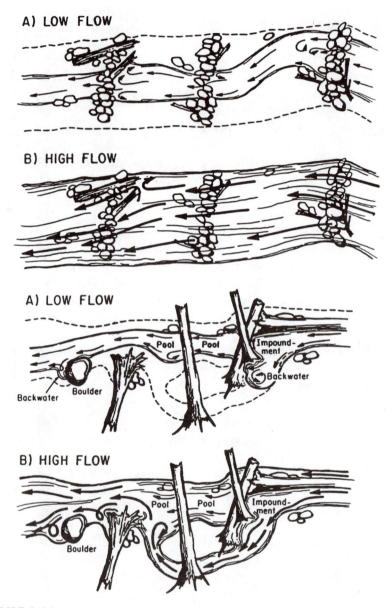

FIGURE 2.35
Stream reaches with and without large driftwood react differently to changes in water flow. In-channel structures, such as boulders, do not increase the number of protected areas available to fish during high water. Large driftwood, however, creates more protected areas during high water. (Adapted from Kaufmann, P.R. 1987. Ph.D. dissertation, Oregon State University, Corvallis.)

FIGURE 2.36
Salmonids use stable driftwood on the floodplain as protective cover during winter floods. (USDA Forest Service photograph by James R. Sedell.)

FIGURE 2.37
Note how the fallen tree (foreground) channels the water against the rootwad (left), which, in turn, buffers the stream bank from the current's velocity. Without the rootwad's buffering effect, however, the current would wash away the stream bank and thereby widen the channel. (USDA Forest Service photograph by James R. Sedell.)

enough high water to move the driftwood on its journey from the forest to the sea. And with each new environment encountered on its journey, the driftwood provides a new habitat in a new location for a new group of plants and animals.

CHAPTER 3

THE MCKENZIE RIVER

The first chapter followed driftwood on its journey from a first-order stream that flowed from a high meadow into the forest, where it met another first-order stream originating in a spring. These first-order streams joined to form a second-order stream, which was joined by another second-order stream to form a third-order stream. The third-order stream did not grow appreciably in size and so it emptied as a third-order stream directly into McKenzie River. The McKenzie River, on the other hand, is a sixth-order stream, which drains all of the forested valleys in its drainage basin between the crest of the Western Cascade Mountains and the Willamette River.

THE STREAM-ORDER CONTINUUM REVISITED

In small streams under the canopy of an aged forest, much of the food for invertebrates comes from leaf litter or wood. As streams become progressively larger and the vegetative canopy along their sides overhangs less and less of the water, the influence of the forest diminishes and the energy basis for the stream is derived more from algae and less from forest litter. The greatest influence of the forest is seen in very small streams, but the greatest diversity of mechanisms for the input of organic foods and of organic habitats occurs in the intermediate (third- to sixth-order) streams.

The direct influence of riparian areas, which is moderate in streams of sixth order and larger, is important. Dense canopies of large old trees provide some shade, and the vegetated riparian zones tend to provide stability to the banks of the main channel. Further, the largest stems of fallen trees in a riparian area may remain in the stream to provide important summer and winter habitats for trout and salmon.

The wide floodplain areas of these larger streams contain complex arrays of side channels, overflow channels, isolated pools, and the greatest abundance of driftwood. Side channels are often created and maintained by large driftwood that has become temporarily anchored and redirects the current. Although the gradient of these large streams is usually less than one percent, rapids and falls may still occur. Alluvial materials (those deposited by flowing water) and driftwood may be deposited initially in quiet areas or river bends, but such accumulations are flushed and rearranged during the flow of high water that follows storms and the rapid melting of snow.

FIGURE 3.1
Driftwood on a gravel bar provides wood to the channel during high water. (Photograph by M. Kellerhals.)

DRIFTWOOD ON THE FLOODPLAIN

Although driftwood at times severely batters streamside vegetation as it journeys seaward, at other times it protects such vegetation, even to the point of allowing the forest to temporarily claim part of the exposed floodplain channel (Figure 3.1).

Well anchored and too large to move, the driftwood poses an obstacle that slows the water's force, causing it to drop part of its suspended sediments on the driftwood's downstream side (Figure 3.2). Here enough soil can accumulate that red alder seedlings become established and grow.

As alders grow, willows, streamside sage, thistles, fireweed, horsetail, sedges, and grasses become established in and around an anchored

A

FIGURE 3.2
(A) Stable driftwood on active floodplain protects trees as they grow and become more flood resistant; (B) red alder growing behind large driftwood; (C) schematic of protected site behind stable driftwood. (USDA Forest Service photographs and schematic.)

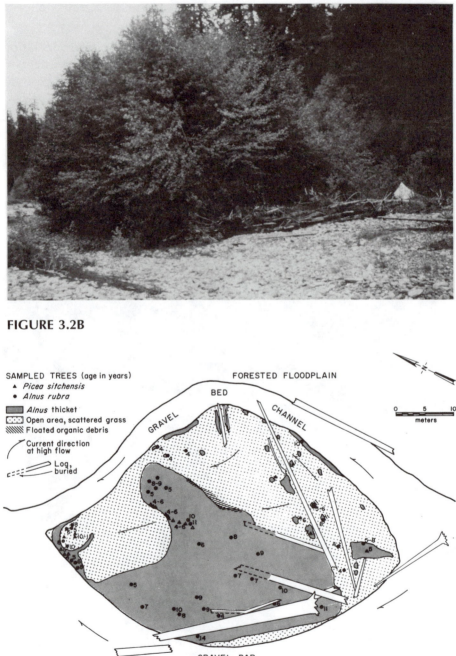

FIGURE 3.2B

FIGURE 3.2C

FIGURE 3.3
Driftwood provides habitat for small mammals. (Photograph courtesy of the USDA Forest Service.)

accumulation of driftwood (Figure 3.3), and small mammals move into the storm-created habitat, including the deer mouse, long-tailed vole, western red-backed vole, Townsend chipmunk, and dusky shrew. In turn, predators of these small mammals, such as long-tailed and short-tailed weasels and mink, begin hunting amongst the accumulations of driftwood on floodplains.

As mice, voles, and chipmunks begin using the accumulation of driftwood as habitat, they simultaneously inoculate the soil with the spores of mycorrhizal fungi, a component of soil necessary for tree growth. The term mycorrhiza (which literally means "fungus-root") denotes the symbiotic relationship between certain fungi, which fruit belowground, and plant roots. Trees, such as pine, fir, spruce, larch, Douglas-fir, hemlock, oak, birch, and alder, depend especially on mycorrhiza-forming fungi for nutrient uptake, a phenomenon traceable back some 400 million years to the earliest known fossils of plant rooting structures.

Mycorrhizal fungi absorb nutrients and water from soil and pass them to a host plant. Fungal hyphae (the "mold" part of the fungus) extend into the soil and act as extensions of the hosts' root systems, since they are physiologically and geometrically more effective in absorbing nutrients and water than are the roots themselves. The host plant, in turn, provides simple sugars to the mycorrhizal fungi, which generally cannot get or make their own food.

Belowgound fungal fruiting bodies or truffles are the initial link between mycorrhizal fungi and small mammals. As a truffle matures, it produces a strong odor, attracting a foraging mammal that digs out and eats the truffle, consuming fungal tissue containing nutrients, water, viable fungal spores, nitrogen-fixing bacteria, and yeast.

Pieces of truffle move to the stomach, where fungal tissue is digested. Undigested material is formed into fecal pellets. A fecal pellet is more than a package of waste products; each pellet contains viable spores of the mycorrhizal fungi, and each fecal pellet also contains the entire nutrient requirement for the nitrogen-fixing bacteria. The yeast, as a part of the nutrient base, has the ability to stimulate both growth and nitrogen fixation in bacteria.

As complicated as this is, it is only a tiny glimpse of the entire complex process, which may in time allow the forest to extend itself onto the floodplain. With time, small mammals and such birds as common mergansers, Harlequin ducks, spotted sandpipers, and nighthawks will use such accumulations of driftwood as habitat for nesting. Other birds, such as song sparrows, violet green swallows, robins, and cedar wax-wings, will use it for perches.

CHAPTER 4

THE WILLAMETTE AND COLUMBIA RIVERS

The Willamette River begins as two primary forks. One heads in the Coast Range southwest of its confluence with the McKenzie River. The other fork rises near the crest of the Cascade Range southeast of its confluence with the McKenzie River. The Willamette is already a large, well-defined stream by the time the waters of the McKenzie River empty into it. Yet its beginnings, as with all rivers, are the myriad small seeps and springs, which give rise to first-order streams that join to become second-order streams, that join to become third-order streams, and so on. Thus, by the time the Willamette reaches the McKenzie River, it is a seventh-order stream. Although hundreds of miles long, it is still fairly shallow with a rocky bottom because it has not yet reached the deep soils of the broad, flat valley that extends northward to the Columbia River.

The Willamette Valley, through which the Willamette River flows, is richly vegetated with grasses, herbs, and savannahs of Oregon white oak. Along the streams and rivers of the valley, however, shrubs and trees were so thick that when the early European explorers encountered this riparian forest they called it "brakes."

The valley bottom itself had few trees because the indigenous peoples routinely burned the valley. By accentuating the boundaries between cover along the waterways and the open areas of forage, this allowed them to hunt large game more easily. Along the upper Willamette River,

Willamette River Corridor Forest
Approximate Area

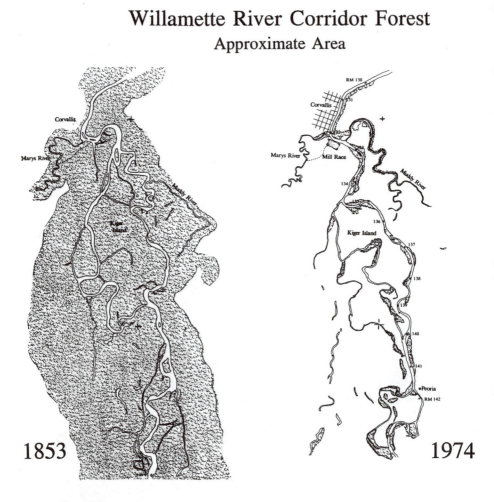

FIGURE 4.1
Contrast of wide, continuous expanse of riparian forest in 1854 with remaining small, fragmented patches in 1974. (Adapted from Habeck (1961) and Johannesen et. al. (1970) by Patricia Benner, Oregon State University.)

however, the riparian forest was one to three miles wide and had a diverse species of trees, ranging from sugar pine on dry, well-drained soils to black cottonwood and Oregon ash in wet areas. Both the width and the interior dampness of the riparian forest made it exceedingly resistant to fire (Figure 4.1).

Debris, which accumulated as a result of floods, provided protected areas in which fine soils were deposited and seedlings grew. The forest was thus a mosaic of ages and species of trees, which could be reasonably predicted based on how wet or dry a site was, how often it flooded, and how long it was submerged by high water. This rich forest was thus in an intimate partnership with the river.

Storms left drifted trees partly submerged, with their rootwads entangled in shrubs and young cottonwood and ash trees along the river's edge where its channel divided into several channels, weaving around islands, sloughs, and a broad expanse of riparian vegetation dominated by Oregon ash and black cottonwood.

In other areas, driftwood was combed out of the flood waters by the trees and dense vegetation of the riparian forest. In still other areas, accumulations of driftwood would lie exposed on gravel bars on the upriver edge of islands or trapped on gravel bars against the river's high, steep banks in the great bends of its channel, where it meandered through the deep soil of the valley floor (Figure 4.2).

FIGURE 4.2
A slow, deep, valley-bottom tributary of the Willamette River, whose waters are heavily laden with soil and whose banks are stabilized by riparian forest. (Photograph by Chris Maser.)

OF FLOODS AND DRIFTWOOD

In a large, multi-channeled river, such as the Willamette throughout its upper valley, relatively small floods can move and store a great deal of driftwood via temporary damming, where some of it blocks the openings of side channels and clogs sloughs. As the river rises, floating driftwood, including whole trees, accumulates at bends in a channel, clogging it, so that water backs up and the turbulent flood cuts a new channel through the valley bottom.

When a new channel is made, more trees topple into the river and are carried downstream. Trees from the adjacent riverside forest and from the forests upstream combine to connect the river with its valley by creating new aquatic habitats and aiding the creation of new oxbow lakes (U-shaped bends in a river that retain water, but which are cut off from the current).

Rivers migrate back and forth across their valleys, cutting off bends, continually eroding some banks only to build others downstream by depositing soil carried in their waters. Streamside forests add to this mutual, physical process by stabilizing banks and sifting flood waters for fertile sediments and driftwood, making floodplains rich areas for farming.

Because big floods generally do not occur often enough to move driftwood rapidly downstream from inland mountains, most driftwood would not even reach the Columbia via the Willamette, let alone the sea, as each piece encountered different circumstances along its journey. Some of the driftwood would become entangled in streamside vegetation; some would be detoured into side channels. Other pieces would become stuck in bottom mud where they eventually became water-logged and sank, remaining for centuries as habitat for fish. The rest would lie partly in or near the water and rot, or become partly water-logged, or both, so that from storm to storm there would be less of it for the water to transport. Thus, driftwood, particularly the smaller pieces, decomposed and was consumed by the river to flow in its current, unrecognizable form, yet vital as energy for the life of the river.

Driftwood that succeeded in entering the Columbia River from the Willamette during winter storms would join other wood, most of which came from adjacent shores and from rivers (Kalama, Lewis, Cowlitz, Washougal, Sandy, Wind, and White Salmon) within about 150 miles of the Columbia's mouth. Some wood, but far less, traveled down other rivers more than 150 miles east of the Columbia's mouth (Deschutes, John Day, Umatilla, Yakima, Snake), and some even from the Columbia itself, which originates in the mountains of British Columbia, Canada. Here, wood from the arterial system of the great Columbia River Basin—some from as far away as British Columbia, Idaho, Wyoming, and Nevada—met, collected, and journeyed to the sea. Some of it stopped short of the ocean, forming such intertidal islands as the Snag Islands and Woody Island along the Woody Island Channel in the Columbia River's estuary (Figure 4.3).

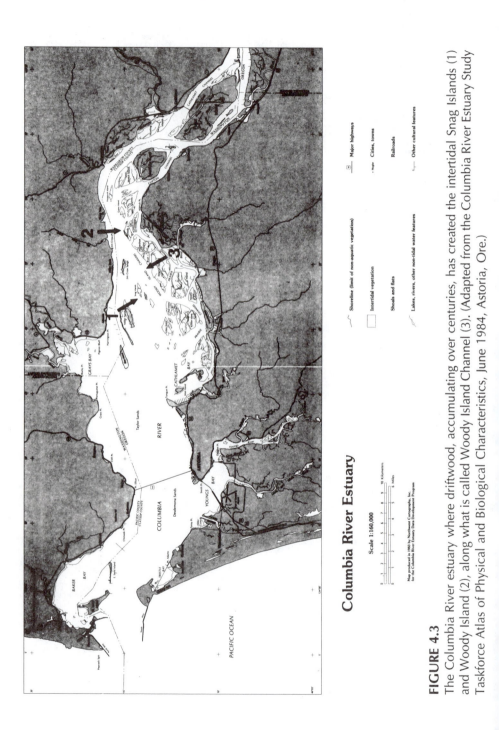

Columbia River Estuary

Scale 1:160,000

Map produced in 1983 by Northwest Cartography, Inc.
for the Columbia River Estuary Data Development Program

Shoreline (limit of non-aquatic vegetation)

Intertidal vegetation

Shoals and flats

Lakes, rivers, other non-tidal water features

Major highways

Cities, towns

Railroads

Other cultural features

FIGURE 4.3

The Columbia River estuary where driftwood, accumulating over centuries, has created the intertidal Snag Islands (1) and Woody Island (2), along what is called Woody Island Channel (3). (Adapted from the Columbia River Estuary Study Taskforce Atlas of Physical and Biological Characteristics, June 1984, Astoria, Ore.)

CHAPTER 5

THE ESTUARY

An estuary is the wide, lower course of a river where its current is met and influenced by the ocean tides. It can also be thought of as an arm of the sea extending inland to meet the mouth of a river.

During the late Pleistocene epoch, some 70,000 years ago, sea level was approximately 325 feet lower than it is today. The Pacific Northwest coast was not glaciated, and rivers had excavated deep valleys in the flat, exposed, coastal plain. Sea levels rose after the close of the Pleistocene, about 10,000 years ago, to submerge the coastal plain and form the present continental shelf.

Pacific Northwest estuaries are at most 15,000 to 10,000 years old. They remain dynamic and changing in physical geography as they migrate across the continental shelf and the present coastal plain in response to changes in sea level, which, in turn, responded to cycles of continental glaciation.

Although the seaward end of many Pacific Northwest river valleys became flooded and marine sediments accumulated in former meanders of the rivers' old floodplains, the main channels remain as the central, deeper channels of modern estuaries, and the estuaries have retained the long, twisting topography of a river in its floodplain valley (Figure 5.1).

Present-day estuaries have accumulated sediments in former channels and on former floodplains. These sediments are usually derived

FIGURE 5.1
Columbia River Estuary between Oregon (bottom of photo) and Washington (top of photo). (1) Columbia River, (2) Willamette River, (3) Clackamas River, (4) Sandy River, (5) Lewis River, and (6) Cowlitz River. (Photographed by the Earth Resources Technology Satellite.)

from silts and muds in the upper reaches of the rivers, which become more sandy near their mouths. Estuarine tidal flats and marshes occupy the former floodplains along the submerged river valleys, which now are estuarine embayments.

HOW ESTUARIES FUNCTION

In the region of Puget Sound in the state of Washington, Pleistocene glaciers cut deep gorges, many of which are interconnected as the extensive marine inlet system. Rivers and streams enter this system and create a shallow layer of fresh to brackish water above the deep marine water at the heads of fjords.

Because neither the flow of the streams and rivers nor the energy of the tide is sufficient to mix the column of water very deeply, the salinity at the bottom of the inlet system remains essentially oceanic, which permits wood-boring marine animals to penetrate far inland. In addition, the extensive tidal flushing and its strong currents in the straits and sounds rapidly move large driftwood, including whole trees, on which the wood-boring marine animals depend.

In coastal portions of Pacific Northwest rivers, the tide affects the water level much farther upstream than the greatest inland penetration of even the most dilute sea water at the highest tide. This hydraulic tidal effect causes the elevations of rivers to rise and fall and can even reverse the direction of a river's flow in the reach referred to as the "tidal river," which forms a barrier to upriver intrusion of salt water. The furthest point upstream that the hydraulic tidal effect reaches, the "head of tide water," is the same throughout the year, irrespective of the river's rate of flow.

The hydraulic tidal effect moves driftwood downstream faster than might be expected because driftwood, stranded on the shoreline by freshets, will constantly be reached by higher high tides and refloated to continue its journey to the sea. The currents created by the outgoing tide also increase the net rates at which driftwood is carried to the sea. In the lower Columbia River, the outgoing surface current can exceed six miles per hour near the river's mouth; only about one fifth of this speed is due to the river's current alone.

The Columbia River, the largest Pacific Northwest river system, is the most striking example of an inland section of tidal river. During low flow, tidal excursions extend over 160 miles upstream, stopped only by Bonneville Dam. Tidal excursions also extend into one of the Columbia's tributaries, the Willamette, at Portland, Oregon.

In the lower 25 to 50 miles of the Columbia River, the tide reverses the surface flow, the extent of which depends on the magnitude of high tides, ranging between 8 and 13 feet at the estuary's mouth. Measurable levels of salinity intrude only about 6 miles upstream during a flood and, rarely, 13 miles upstream during periods of summer low flow.

The difference in density between fresh water entering an estuary from a river and sea water entering from the mouth of an estuary can result in vertical stratification of the mass of estuarine water. In a density-stratified estuary, the less dense, low-salinity surface layer of mixed fresh water and sea water flows downstream toward the sea, and the denser, more saline sea water, originating at the mouth of the estuary, flows upstream near the bottom.

Sea water, flowing upstream along the bottom, becomes mixed upward into the downstream flow of fresh water along the estuary's entire length. Upward mixing continually increases both volume and salinity of the upper layer as it flows down the estuary; the closer to the sea the downstream flow gets, the denser and more saline it becomes.

The presence and extent of vertical stratification of water density vary seasonally in Pacific Northwest estuaries. In the lower estuary there is a strong upstream flow of saline, oceanic water near the bottom and a downstream flow of dilute, mostly riverine water at the surface. The near-bottom upstream flow of oceanic water gradually weakens as it moves up the estuary and finally ceases at some point upstream in mid-estuary, above which water flow is downstream both near the bottom and at the surface.

The point in mid-estuary where water flow near the bottom is not appreciable, either upstream or downstream, is termed the "null point" (Figure 5.2). With no appreciable upstream or downstream movement of water at the null point, water-logged driftwood of all sizes is simply rolled around on the bottom.

The position of the null point varies with the volume of water discharged by the river and is thus closer to the river's mouth during the rainy winter season, when the downstream flow of fresh water dominates physical conditions in the estuarine water. Below the null point, oceanic water and its sediments may be transported upstream along the estuary's bottom. At and above the null point, there is a region of increased turbidity and sedimentation (the area of maximum turbidity), where suspended sediments in the riverine water enter a region of decreased velocity and cloudiness (Figure 5.3).

Water-logged driftwood may become incorporated into the sediments in the area of maximum turbidity, only to be periodically remobilized and flushed toward the estuary's mouth during winter floods.

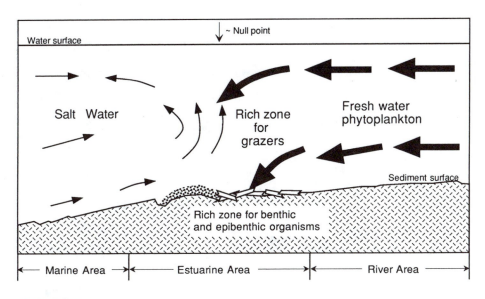

FIGURE 5.2
Diagram depicting the flow of water on the bottom of the estuary moving upstream from the ocean at high tide and downstream from the river. The region of the mid-estuary where the near-bottom flow of water is negligible in either direction is the null point and also the area of maximum turbidity. Driftwood often is incorporated into the sediments of the bottom of the channels of estuaries in this zone. (Adapted from the Columbia River Estuary Study Taskforce Atlas of Physical and Biological Characteristics, June 1984, Astoria, Ore.)

Floating driftwood, on the other hand, tends to be retained in the upper estuary in summer, when the influence of incoming fresh water is minimal. During summer, therefore, driftwood is moved downstream only during tidal cycles sufficiently high to reach it and float it toward the sea, where much of it becomes grounded and accumulates on the deltas of such major rivers as the Willamette.

Pacific Northwest estuaries typically have more tidelands than subtidal areas. The shorter and broader an estuary, the faster a particular existing mass of estuarine water is flushed into the sea, assuming unrestricted downstream flow. In summer, for example, such an estuary can be flushed in less than ten days. On the other hand, the upper regions of an estuary with more tidelands than subtidal areas, which are long, twisting, and often relatively narrow compared with the breadth of its mouth, may take two to four times as long to flush.

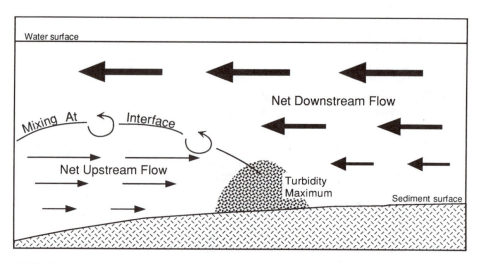

FIGURE 5.3

Schematic representation of fresh water encountering saline water at the summer season of lowest river flow. The area of maximum turbidity concentrates dying and dead plankton and fine, organic materials, creating a rich feeding zone for estuarine organisms. (Adapted from the Columbia River Estuary Study Taskforce Atlas of Physical and Biological Characteristics, June 1984, Astoria, Ore.)

Driftwood tends to be retained longer in upper regions of estuaries with the longest flushing times, while floating driftwood in the lower segments of estuaries exits more rapidly. Marine wood-boring organisms adapted to the annual regime of salinity and temperature of the upper region of an estuary grow and reproduce rapidly in the warm temperatures reached above the summer null point. Decreased flushing increases the chances that dispersal stages of marine wood-borers will remain longer in the upper estuary, which increases their chances of encountering new driftwood, also retained longer in the upper estuary.

Just off the mouth of the Columbia River is a tidally related, clockwise-rotating eddy displaced slightly to the north. The eddy is strongest at ebb tide and brings driftwood onshore just north of the river's mouth. The large rafts of drifting trees off the mouth of the Columbia, noted in the historical records of early sea explorers, were probably related to this eddy. Such a system collects driftwood washed out of the Columbia and holds it in nearshore ocean currents.

The fate of driftwood entering an estuary or the open ocean is largely determined by the marine organisms that cause its biological deterioration and incorporation into the energy flow of the marine ecosystem. A comparison of the kinds of organisms involved in and their relative roles in marine, freshwater, and terrestrial environments reveals many contrasts.

Unlike the fate of fallen trees in terrestrial and freshwater environments, wood in estuarine and marine environments is rapidly attacked by animals long before enough fungal rot takes place to soften the wood. In addition, wood in estuarine and marine environments is dispersed before, not after, the initial microbial decomposition of its constituents.

Marine organisms that degrade wood entering an estuary or the sea belong to different taxonomic groups than those dominating this process in freshwater and terrestrial habitats. Marine wood-degrading organisms are not as diverse as their fresh water and terrestrial counterparts, and the relative roles of microorganisms and animals differ in the initial attack on materials that contain cellulose.

Marine fungi and bacteria use lignin and cellulose as sources of energy, but they appear to play only minor roles in the initial invasion and degradation of wood in marine habitats. The major biodegradation of wood in marine habitats is caused by a few genera of wood-boring crustacea and a larger number of wood-boring bivalve molluscs.

Marine fungi, such as Fungi Imperfecti and Ascomycetes, help to decompose cellulose, but their ability to break it down rapidly or extensively in large pieces of driftwood appears to be much less pronounced than that of their terrestrial counterparts. Most of the fungi, which break down lignin, attack only the surface layers of the wood, and then no deeper than about one sixteenth of an inch.

Members of the marine fungal genus *Luworthia*, however, penetrate deeply into the wood and are more comparable in their ecological function to terrestrial wood-decaying fungi. Unlike driftwood in fresh water, the presence of marine fungi in waters of the Pacific Northwest is not a prerequisite for the initial attack on the wood by the primary marine wood-boring organisms, namely gribbles and shipworms.

The role of marine cellulose-digesting bacteria is largely obscure, but many species, which can actively break down cellulose, have been identified. Aerobic bacteria, which can decompose cellulose, are abundant in sea water and are widely distributed in the marine sediments. These organisms must play an important role as remineralizers of cellulose, which contains organic material, but their exact role in marine and estuarine cycles of carbon is unknown.

Some organisms cause rapid deterioration of cotton nets in sea water; others colonize and attack submerged driftwood. The direct attack, either previously or concurrently, on wood by bacteria is not a prerequisite to activity by marine wood-boring organisms.

GRIBBLES

Gribbles are wood-boring isopod crustaceans that decompose wood in estuarine and marine waters of the Pacific Northwest (Figure 5.4). Only two species are involved: the endemic northern gribble and the southern gribble, found in upper segments of bays and estuaries, which was possibly introduced to the region. The northern gribble, a cold-water species, occurs in bays and the lower segments of estuaries, as well as in reproducing populations in the cold water of the outer coast. The southern gribble, on the other hand, can survive in the cold water of the outer coast but can neither reproduce nor establish new colonies.

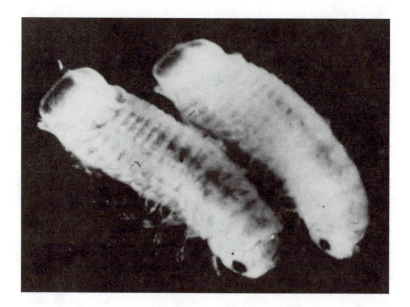

FIGURE 5.4
Gribbles. Their heads are in the lower right of the photograph. (Forest Research Laboratory, Oregon State University, photograph courtesy of Jeffery Morrell.)

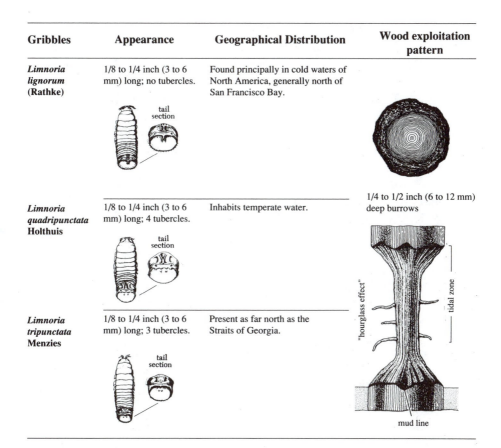

Gribbles	Appearance	Geographical Distribution	Wood exploitation pattern
Limnoria lignorum (Rathke)	1/8 to 1/4 inch (3 to 6 mm) long; no tubercles.	Found principally in cold waters of North America, generally north of San Francisco Bay.	
Limnoria quadripunctata Holthuis	1/8 to 1/4 inch (3 to 6 mm) long; 4 tubercles.	Inhabits temperate water.	1/4 to 1/2 inch (6 to 12 mm) deep burrows
Limnoria tripunctata Menzies	1/8 to 1/4 inch (3 to 6 mm) long; 3 tubercles.	Present as far north as the Straits of Georgia.	

FIGURE 5.5

Gribbles: physical characteristics, geographical distributions, and patterns of exploiting wood. (Reprinted from Helsing, G. *Recognizing and Controlling Marine Wood Borers,* Publication SG 49, Oregon State University Extension Service, Corvallis, revised 1981.)

Both species resemble small, elongated versions of the pill bugs of terrestrial environments. They live for one or two years within the protective confines of their bored tunnels. The northern gribble is the larger species; females may be nearly 0.02 of an inch in length, whereas female southern gribbles reach only 0.012 of an inch (Figure 5.5).

Both gribbles behave similarly. Although their abundance and wide distribution make gribbles major wood-boring organisms, they apparently cannot detect the presence of wood from a distance, but are quick to identify it on contact. Dispersal along the outer coast or between bays must therefore be mainly as colonies in floating wood rafted about the sea.

Pacific Northwest gribbles are predominantly animals of the intertidal or shallow, subtidal areas. Although they readily bore into any submerged wooden surface, they attack mostly wood found in the lower intertidal zone or the mid-tidal subsurface portions of such large wood as pilings (Figure 5.5).

Active borers, they form dense colonies of long, interconnecting tunnels the diameter of their bodies just below the surface of the wood. Such colonies can cover entire surfaces of whole drifted trees.

As a colony burrows deeper, the outer region of the wood, riddled with the interconnected older tunnels, becomes water-logged and softened by microorganisms. Consequently, it is fragile and easily broken into small pieces by abrasion and wave action, which gradually remove the outer layers no longer occupied by the colony (Figure 5.6).

Gribbles burrow continuously, removing and consuming small pieces of fresh, intact wood fiber from the blind ends of their tunnels. Fibers thus removed are further reduced in size with the rasp-and-file mechanism of their mandibles, and the resulting wood powder is then ingested.

Gribbles ingest all of the wood they remove during their burrowing activities. Approximately 45 percent of the ingested wood is utilized; the rest is voided as fecal pellets composed of finely ground wood fibers, lightly bound together in a short, flattened, cylindrical form by a thin membrane of chitin. These fecal pellets are continually washed from the gribbles' tunnels by water currents created as gribbles ventilate their tunnels for respiration. Although the pellets sink rapidly in calm water, they are light enough to be easily suspended and dispersed by wind-driven turbulence and the currents of shallow estuarine and coastal waters.

Ecologically, the fecal pellets function as a source of carbon, some of which is deposited on the surface of estuarine sediments while the rest is transported to the open ocean. In quiet waters, small piles of fecal pellets accumulate under the colonized driftwood.

Gribbles transfer fine particles of wood to the pool of carbon on bottom sediments by converting trees directly into non-buoyant wood powder that enormously increases the surface area of the wood. Additional microbial decomposition of the fecal pellets by cellulose-digesting

FIGURE 5.6
Gribbles weaken wood by feeding. Waves wash away weakened wood, causing hourglass shape. (Forest Research Laboratory, Oregon State University, photograph courtesy of Jeffery Morrell.)

bacteria on the surface of the sediments may channel this source of carbon into the detrital food web of bottom dwellers.

Organic detritus is the principal energy source for estuarine and shallow-water marine-bottom food webs. A major source of this detrital carbon is often the waste products of such animals as gribbles. Some of the food web of the open ocean is also based on detritus, and the fecal pellets of gribbles are an appropriate size for direct ingestion by zooplankton, which are small, free-floating invertebrates.

Although gribbles remove a significant fraction of the total weight of the wood that passes through their digestive tracts, they do not use all components with equal efficiency. Nevertheless, in one step, a gribble converts a piece of driftwood, such as an old Douglas-fir tree, from its original, massive form into microscopic, non-buoyant particles that are freely dispersed in estuarine and marine waters and are deposited in estuarine and marine sediments.

SHIPWORMS

Other than gribbles, the only shallow-water, marine, wood-boring animals present in Pacific Northwest waters are shipworms, which came to be considered pests because they infested the wooden hulls of the ships of early voyagers and sailors. Even Captain Cook feared shipworms. He did not trust the new procedure of fitting a copper sheathing over the outside of a ship's hull, but preferred the old reliable double-hull method of protection in which the space between the outer, wooden sheathing and the true hull was filled with animal hair and tar.

There are two species of shipworms in the Pacific Northwest, one native and one introduced (Figure 5.7). They are the most important wood-boring organisms in the estuarine and shallow marine waters of the region.

The introduced shipworm likely arrived in the hulls of ships sailed from tropical waters to the West Coast in the 1500s. From the late 1600s to the end of the 1700s, regular and widely used trade routes had been established between Asia, Polynesia, and the west coast of North America. Shipworms and gribbles, as well as other organisms—both plants and animals (termed "fouling agents")—were transported and introduced worldwide in wooden-hulled ships.

The introduced shipworm is a warm-water species of restricted distribution in the Pacific Northwest—in San Francisco Bay, California, the inner parts of a few bays on outer Vancouver Island, Canada, and, as of 1981, Willapa Bay, Washington. It can reach five inches in length and two tenths of an inch in diameter in its ten-week life span. It is more resistant to periods of lowered salinity and elevated temperature than is the native shipworm. It can thus establish populations in the upper estuarine areas, which the native shipworm cannot colonize.

The native shipworm of the North Pacific is found in the open coastal

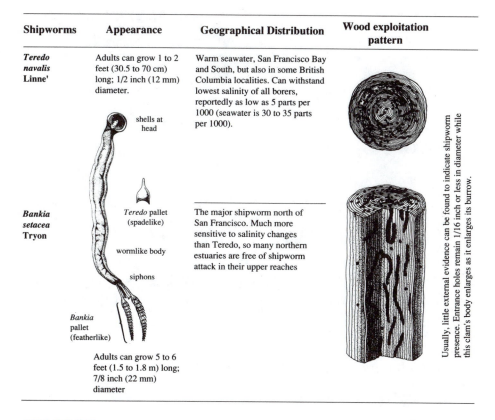

Shipworms	Appearance	Geographical Distribution	Wood exploitation pattern
Teredo navalis Linne'	Adults can grow 1 to 2 feet (30.5 to 70 cm) long; 1/2 inch (12 mm) diameter. shells at head *Teredo* pallet (spadelike) wormlike body siphons	Warm seawater, San Francisco Bay and South, but also in some British Columbia localities. Can withstand lowest salinity of all borers, reportedly as low as 5 parts per 1000 (seawater is 30 to 35 parts per 1000).	Usually, little external evidence can be found to indicate shipworm presence. Entrance holes remain 1/16 inch or less in diameter while this clam's body enlarges as it enlarges its burrow.
Bankia setacea Tryon	*Bankia* pallet (featherlike) Adults can grow 5 to 6 feet (1.5 to 1.8 m) long; 7/8 inch (22 mm) diameter	The major shipworm north of San Francisco. Much more sensitive to salinity changes than Teredo, so many northern estuaries are free of shipworm attack in their upper reaches	

FIGURE 5.7

Shipworm: physical characteristics, geographical distributions, and patterns of exploiting wood. (Reprinted from Helsing, G. *Recognizing and Controlling Marine Wood Borers,* Publication SG 49, Oregon State University Extension Service, Corvallis, revised 1981.)

ocean as well as in bays and lower parts of estuaries. Because of its abundance, large size, and rapid growth, it is the major marine wood-destroying animal in these waters.

A shipworm is a specialized, bivalved (two shell halves hinged together) mollusc of the family Teredinidae. It has an elongated body and uses toothed ridges on its reduced, specialized shell for mechanically rasping its way into wood (Figures 5.8 to 5.10).

A shipworm secretes a calcium-carbonate lining in its tunnel while boring through the wood. As the animal grows, most of its body lies

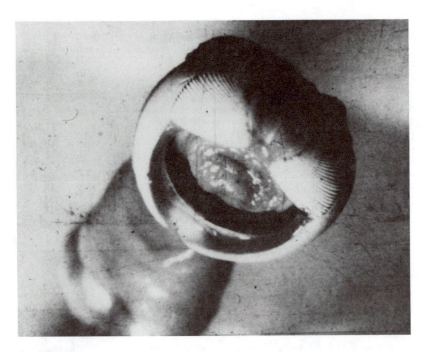

FIGURE 5.8
The head of a shipworm with the pair of rasping shells that aid in boring the wood. Note the elongated, worm-like body that extends into the background. (Forest Research Laboratory, Oregon State University, photograph courtesy of Jeffery Morrell.)

FIGURE 5.9
A shipworm's rasping shell position in its tunnel in the way it is used to bore through the wood. (Photograph by Ruth Turner.)

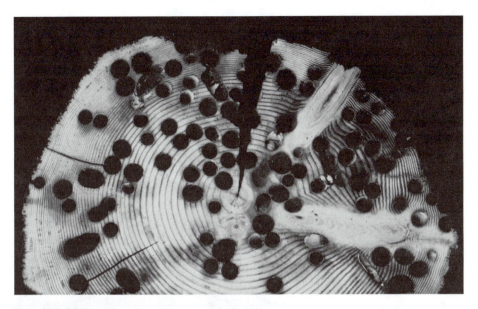

FIGURE 5.10
Cross-section of wood with numerous tunnels of shipworms in the inner portion; note the white, calcified linings of several tunnels. The somewhat frayed surface of the wood was exploited by gribbles. (Forest Research Laboratory, Oregon State University, photograph courtesy of Jeffery Morrell.)

within the older part of the tunnel, protected by the calcified lining (Figure 5.11). From this position, a shipworm's tube-like siphons extend into the surrounding water through the original point of entry into the wood (Figure 5.7).

The portion of its body bearing the shell is at the opposite end from the siphon. The tunnel is unlined here and is where the shipworm burrows. Shipworms bore continuously, lengthening the tunnel and increasing its diameter as the individuals grow. They stop boring only when the space within the wood is exhausted, at which time they seal off the blind end of their tunnels with a calcareous lining, which joins the preexisting tunnel lining (Figure 5.11). Shipworms have life spans of only one or two years, but they bore rapidly and thus have a tremendous impact on driftwood.

The native shipworm breeds when the sea-water temperature is relatively low. The most commonly observed pattern of reproduction in the Pacific Northwest is an intense period of larval settlement from

FIGURE 5.11
An X-ray image of wood showing calcified linings of the shipworms' tunnels and shells used in boring. (Photograph by Ruth Turner.)

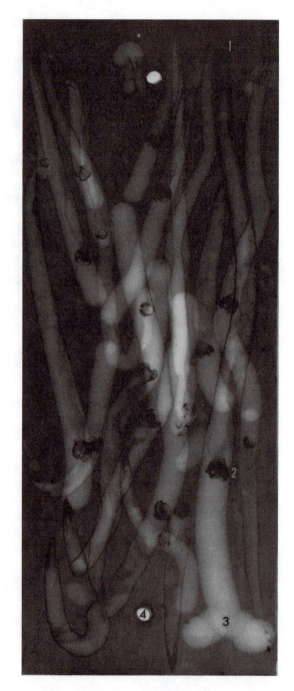

September to December, which corresponds to the onset of winter storms bringing fresh driftwood down coastal rivers to the sea.

Shipworms infest wood at virtually any time because larval settlement occurs throughout the year. Continual larval settlement is important, because coastal rivers derive their peak flows from winter storms, whereas interior rivers, such as the Columbia and Snake, derive their peak flows from spring snowmelt in the mountains.

The native shipworm lays eggs that develop into free-floating larvae, which feed on tiny plants and animals floating with them. A long larval phase of three to four weeks permits larvae to be extensively dispersed in the coastal currents of the Pacific Northwest.

A larval native shipworm first attaches itself to a piece of driftwood and then quickly penetrates, using cutting teeth that develop on its shell after it becomes attached. The metamorphosed larva begins boring into the wood, completely burying itself within 24 hours.

Individual native shipworms that settle on suitable driftwood in December reach about eight inches in length by April. In the warmer waters of April through July, they may grow from 8 to 20 inches in length.

In the waters of Washington State, the native shipworm can become nearly an inch in diameter and about three feet long. In San Francisco Bay, individual shipworms grow to sexual maturity within a month after settling on suitable driftwood.

The life span of an adult native shipworm in Pacific Northwest waters is at least seven months, during which time the largest, most rapidly growing individuals can, in five months, remove up to 16 inches of wood. Growing some two feet in six months, infestations of native shipworms are often so dense that their tubes nearly touch one another, quickly converting the interior of large, solid drifted trees into a finely divided powder, which in turn provides food for filter feeders as it disperses into the estuarine environment.

Because the native shipworm must bore to grow, much of the wood is not digested but rather is flushed directly out of the tunnel as fecal pellets. The fine particles of wood sink in quiet water but are easily suspended in turbulent water. The powdered wood produced by the native shipworm, like that produced by gribbles, becomes fine detritus, which directly enters the detrital-based food web occupied by flat worms, round worms, predatory snails, and other invertebrates.

The native shipworm is highly opportunistic with rapid growth, relatively short adult life span, high rate of reproduction, and long-lived pelagic larvae, which are widely dispersed and capable of rapidly colonizing new driftwood. Native shipworm infestations are intense in the coastal waters of the Pacific Northwest, especially in the harbor at Vancouver, British Columbia, Canada, where some new settlement of larvae occurs in almost every month of the year.

Native shipworms, like gribbles, use the cellulose fraction of the wood they eat, and by the time the ingested wood becomes fecal pellets, an average of about 58 percent of the total carbohydrate content has been removed. The lignins and the remaining cellulose, which are not digested, are voided as finely divided particles in fecal pellets and form part of the detrital energy base of the estuarine and marine environments.

DRIFTWOOD IN ESTUARIES

Winter floods drift large trees downstream into the tidally influenced regions of such coastal rivers as the Siletz, Umpqua, and Coquille and their estuaries (Figures 5.12 to 5.14). Tide influences the extreme upper region of these estuaries, but water salinity is too low for the survival of

FIGURE 5.12
A great flood can supply beaches with enough driftwood to last years. (Photograph by Lewis Nelson.)

FIGURE 5.13
Whereas floods carry driftwood from the land to the sea, winter storms and seismic waves deposit driftwood at the extreme upper edges of beaches, estuaries, and salt marshes. (Photograph by Neal Maine.)

wood-boring estuarine animals. The ecological functions of the drifted trees in this part of the estuary are therefore the same as those upstream from the tidal segment.

Drifted trees influence the extreme upper estuary mainly by their large mass, forming heavy, solid objects with a firm surface in an environment where the bottom is primarily fine sediment. In the tidal segment of coastal streams and rivers, they create riffles and provide shelter from predators for fish, such as stickleback, sturgeon, starry flounder, and juvenile and adult trout and salmon.

Drifted trees can also affect local water flow by creating turbulence and thereby affecting the pattern of sedimentation and formation of bars and mudbanks. Emergent parts of drifted trees stranded in the channel or partly or wholly stranded on tidally exposed banks are used as protected perches by water birds during their daily cycles of rest or are used

FIGURE 5.14
Driftwood quickly becomes incorporated into high salt marshes. (Photograph by Chris Maser.)

as perches from which to hunt by such predatory birds as great blue herons (Figure 5.15) and bald eagles.

Biological decay of driftwood in the uppermost segment of an estuary is slow, taking many decades, especially for trees that are partly embedded or buried in typically anaerobic, upper-estuarine muddy sediments. Under such conditions, only cellulose-digesting fungi are available as agents of decomposition; no animals are present to decompose wood.

Battering by large driftwood rafted downstream on flood waters or being uplifted by such flood waters and carried by tides is one of the major forces that breaks apart drifted trees that have become stranded in the upper estuary. Drifted trees rolled by flood waters are caught on tidal flats by stubs of broken branches that become embedded in the sediments.

Floods and tides eventually move more driftwood downstream into the upper estuary, the brackish, tidally influenced region above the summer null point. Salinity falls drastically in winter, and most marine ani-

FIGURE 5.15
Great blue heron resting on driftwood at low tide. (Photograph by Neal Maine.)

mals, not adapted to prolonged periods of lowered salinity, cannot survive. Because of the reduced flow of water in summer and the limited tidal flushing, water temperature in the upper estuary is significantly higher than in the lower segments of the estuary. Further, some estuarine wood-borers are adapted to live above the summer null point.

The lower segment of the estuary, from the null point to the mouth, is dominated by the marine influence. Its salinities and temperatures are close to that of the open ocean, yet drifted trees and other large driftwood have the same physical role in this part of the estuary that they do in the upper estuary. In contrast with the upper estuary, however, wood in the lower estuary disappears faster through the activities of both the northern gribble and the native shipworm.

Between the two active species of wood-borers, any wood remaining in the lower estuary is destroyed in about a decade, simultaneously attacked from both the outer surface and the interior. The rate of destruction is dramatically greater than that in the upper estuary, which may be about two decades, because there the southern gribble alone is responsible for the initial degradation of wood.

Both the mass and the physical properties of the driftwood are altered. Water-logging increases as the wood is converted into a friable, riddled mass, which is rapidly abraded by the action of waves and by the battering of other driftwood. The result, again, is the direct alteration of large masses of wood into smaller and smaller pieces through abrasion and into a fine, non-buoyant powder through the activities of wood-borers. Most of the powder is introduced directly into the detritus-based food web of the estuary, some of which will be borne to the ocean as a source of energy.

HOW SOME ANIMALS USE DRIFTWOOD

Many of the drifted trees and other pieces of driftwood are used for a time during their journey through an estuary by a variety of vertebrates. Bald eagles, for example, use large driftwood located on tidal flats and alongside estuarine channels away from the shore. Although they neither perch near the shore because of the proximity of people nor sit directly on the soft mud, they will perch on driftwood exposed on intertidal flats at low tide, even if the perches are only slightly elevated above the mud surface.

Eagles forage from a central place, and thus the use of perches near shore is energy-efficient. They perch in places providing good views of both water in open channels, where aquatic birds congregate, and tidal flats, where they watch for opportunities to capture aquatic birds or to steal prey caught by gulls and herons. In addition, because centrally located perches provide greater potential for immediate access to sources of food, the time spent flying to and from foraging areas is reduced.

Green herons, great blue herons, and great egrets are predators that prefer to stand in water waiting for prey to pass by. Although they can wade in water up to eight inches deep, they use as perches any large driftwood that remains partly out of the water at high tide.

When perches on tidal flats or in channels are not available, great blue herons must alternatively perch in distant trees, resting during high tide, and hunt by wading during low tide. Those herons fortunate enough to establish territories on floating but anchored driftwood slowly patrol along the edges of the wood at both low and high tide, looking for fish.

These individuals are able to avoid long flights to and from foraging areas and require smaller feeding territories because the rafts of driftwood are over water deeper than herons can wade.

Great blue herons will defend feeding territories on rafts of driftwood and confine virtually all of their feeding activity to these territories, increasing their energy efficiency. Driftwood rafts give herons access to larger fish than they could catch by wading on the shallow tidal flats because small fish tend to aggregate in the deeper water around the wood, which in turn attracts larger fish. Herons also establish and defend foraging territories on tidal flats containing such suitable perches as anchored, stationary drifted trees.

Cormorants and brown pelicans require perches for resting between periods of foraging, and they prefer perches surrounded by water and away from shorelines. Consequently, they use large pieces of stranded driftwood when available. Small shorebirds, which feed on tidal flats at low tide, also require perches for resting during high tide and use any available, emergent wooden object.

There is also an indirect relationship between drifted trees in estuaries and the marine birds that feed on eggs of the Pacific herring. Herring, entering estuaries to spawn, require such solid objects as driftwood in the intertidal area on which to fasten their eggs. Masses of herring eggs are attached to estuarine algae, rocks, shells, pebbles on the tidal flats, and on any available driftwood, especially the branches of drifted trees stranded in the estuary.

Using the preference of Pacific herring for finely divided surfaces on which to fasten their eggs, the aboriginal peoples placed shrubs and cut branches in the estuaries during the spawning season to collect the herring's eggs for food. Currently, there is a drastic shortage of suitable spawning areas in Oregon estuaries, and the herring's eggs are laid in overlapping, crowded masses.

Crowding greatly reduces successful hatching by reducing the oxygen supply within the egg masses. In addition, masses of excess, loosely attached eggs break free, are rolled across the tidal flats by waves, and accumulate on the beaches. Both the beached eggs and those that remain attached in place are readily eaten by birds, including such non-aquatic species as the American crow. Gulls alone can consume two thirds of the eggs on a beach at low tide.

During the Pacific herring's spawning season in Yaquina Bay, Oregon, the total resident bird population increases from about 1000 or

2000 individuals of all species to about 10,000. Thus, the limiting factor in the successful spawning of herring—the extent of suitable egg attachment sites, such as stranded trees with branches in the intertidal area—is also a limiting factor in the availability of the herring's eggs as food for birds associated with the estuaries.

Estuaries are also important for smolts of such fishes as coho, chinook, chum salmon, and sea-run cutthroat trout, which use estuaries as sites for completing their adaptation to sea water prior to entering the open ocean. While in the estuarine environment, which may be about two months, smolts aggregate around large, stable, in-stream driftwood, which affords structurally complex cover, low light intensity, and slow water velocity. Shelter from swift water during spring freshets and tidal flushing is important to prevent smolts from being washed to sea before completing the smoltification process, especially because they are exceedingly vulnerable at this time. Some of this critical habitat is today provided by docks, log pilings, and log rafts.

Mammals also use driftwood. Harbor seals use large driftwood, either in rafts or as stranded trees away from shore and disturbance by humans, as areas to climb out of the water and rest, termed "hauling out" (Figure 5.16). Harbor seals, the seals most commonly seen near shore and in bays and estuaries, haul out only where water is constantly accessible:

FIGURE 5.16
Harbor seal resting on driftwood. (Photograph by Neal Maine.)

FIGURE 5.17
Spotted skunks use driftwood as home. (Reprinted with permission from Mead, R.A. and K.B. Eik-Nes. 1969. *J. Reprod. Fert.* Suppl. 6, 397–403.)

on tidal sand bars, tidal mud flats, small rocks, islands, reefs, and large driftwood. The availability of suitable areas allows the seals to haul out at any time; otherwise they come out only in the absence of people at night.

Large driftwood partly or mostly buried in sand dunes is used by spotted skunks (Figure 5.17) as preferred sites under which to construct their dens.

Smaller pieces of driftwood, such as the partial stem of a red alder cut by a beaver upriver (Figure 5.18), are too big and heavy for the wind to move. If the downwind end of such wood becomes partly buried in the sand, it is further stabilized, and a tiny, elongated pile of sand, formed on the lee side of the wood, anchors it even more. On the protected, inland side of the wood, the coast tiger beetle seeks shelter from wind-blown

FIGURE 5.18
Following winter storms, coastal beaches acquire sticks of red alder or other trees cut and stripped of bark by beavers living in streams and rivers. (Photograph by Chris Maser.)

sand in the small pocket of relatively quiet air formed by the physical barrier of the wood and the tiny sand dune.

DRIFTWOOD AND SALT MARSHES

Salt marshes are densely vegetated, low coastal wetlands at elevations within the annual, vertical range of regular tidal fluctuations (Figure 5.19). Plants of the salt-marsh community are of terrestrial origin and are capable of growing in saturated estuarine sediments and of withstanding the stresses of both salinity and tidal inundation. Salt marshes have high annual rates of vegetative production of plants, a significant portion of which is exported to the estuary as detritus.

Driftwood in the upper estuary, transported downstream in flooding rivers, often in concert with high winter tides, is deposited on the highest

TIDAL RIVER & ESTUARY CHANNELS

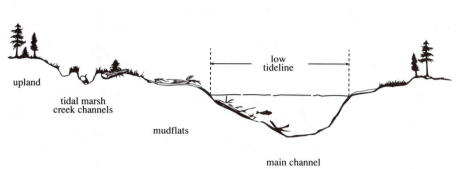

FIGURE 5.19
Cross-section of an estuary showing the relationship of the salt marsh to adjacent forest, mud flats, and estuarine channel.

point reached by the highest water. This usually results in a dramatic shoreline of large driftwood, sometimes in continuous rafts in the high salt marsh, which is the characteristic plant community at this level of tidal influence within the upper estuary. Here large driftwood functions as an agent of disturbance in an otherwise dense, productive, and uniform plant community.

The requisite conditions for salt marshes are lowered salinity, extensive areas of soft sediments at the level of high tide, and low wave energy. Such conditions are virtually restricted to estuaries in the Pacific Northwest, where there are no true, open coastal salt marshes.

Salt marshes are important components of estuaries in the Pacific Northwest. Despite their relatively high elevation at the upper limits of tidal influence, salt marshes function as buffers, muting effects of floods and storms by allowing turbulent waters to dissipate much of their energy over the marsh's vast expanse. Further, the tidal creeks draining salt marshes are extensively used at high tide by migratory waterfowl and juvenile fishes, including salmonids.

Driftwood is abundant in an upper salt marsh, and the boundary of the marsh is clearly delineated by extensive and continuous piles of large driftwood, including whole trees. Large driftwood is also scattered throughout the marsh and remains in place for long periods, which allows the general level of the marsh around them to increase through the deposition of silt and the accumulation of organic matter.

When these trees are refloated during unusually high tides, floods, or coastal storms, shallow depressions remain in the marsh's sediment, reminders that drifted trees had rested there. Refloated driftwood is an important source of disturbance, creating habitat diversity and therefore succession within and distribution among marsh-plant communities within the high salt marsh.

These depressions increase habitat diversity and hold water at low tide in summer, which harbors juvenile fishes. Depressions left by refloated driftwood may enlarge and remain unvegetated for long periods because of increased salt concentration during summer, when the sediment surface dries or oxygen levels are low due to stagnant soil water.

Further, the relative mobility of large driftwood in different parts of the high salt marsh influences the successional processes of the plant communities in opposite ways. For example, one of the highest portions of the marsh is dominated by a wetland-forest community of Sitka spruce, red alder, and willow, which slowly invades the marsh as the progradation of river-borne sediments accumulates. Progradation (the opposite of degradation) is the building of land upward and outward through sediment deposition.

This forest advances in some places by actively colonizing the stable piles of large driftwood left by past storms. In other parts of the marsh, which are exposed to the waves of winter storms, the edge of this same forest community retreats, often because large driftwood batters against the trees on the upper shore of the marsh during abnormally high tides.

In addition to the forest itself, large driftwood embedded in the marsh is colonized by terrestrial plants unable to grow directly on the marsh soil because of its salinity or lack of oxygen. In fact, most of the spruce growing in the high marsh away from the forest's edge are growing on "nurse trees," with but few of their roots extending into the soil of the marsh. A nurse tree is a drifted tree that is rotten enough to form a good rooting medium for other plants (Figure 5.20). Spruce trees can live several decades on driftwood nurse trees.

Driftwood in the upper marsh is exposed to a wet, predominantly freshwater, environment. Large driftwood typically resembles fallen, decaying trees in the terrestrial portion of the forest rather than trees elsewhere in the estuary, which are attacked by estuarine wood-borers. If these trees remain in place in the high marsh, they are attacked by wood-decaying fungi and show the typical sequence of internal rotting exhib-

FIGURE 5.20
Sitka spruce growing on driftwood in Oregon coastal marsh. (Photograph by Chris Maser.)

ited by trees on the forest floor. (This is described in *Forest Primeval, The Natural History of an Ancient Forest,* by C. Maser.)

Since European settlement, which began about 1792 and was flourishing by 1800, estuarine marshlands of the Pacific Northwest have changed greatly. On the one hand, salt marshes have been put in a straitjacket of dikes along the rivers to disconnect them from the estuaries and simultaneously have been ditched to drain them quickly of water. The extensive degradation and removal of marshlands from the estuarine ecosystem has been caused by the building of dikes and drainage ditches to gain land for agriculture and pastures for dairy cattle, by road building, and by filling in salt marshes to gain higher ground for construction of housing developments and shopping centers (Figure 5.21).

On the other hand, the rapid progradation (the building outward toward the sea) of the salt marshes can be a result of such human activities as logging, road construction, and cultivation of the land, all of which

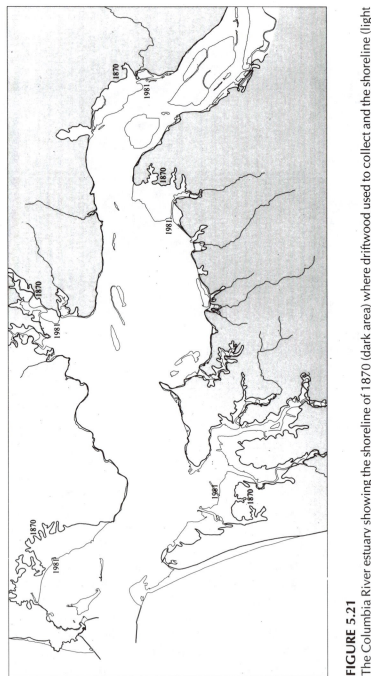

FIGURE 5.21

The Columbia River estuary showing the shoreline of 1870 (dark area) where driftwood used to collect and the shoreline (light area) as of 1981. Note how many of the original intertidal areas have been filled in during the intervening 111 years. (Adapted from the Columbia River Estuary Study Taskforce Atlas of Physical and Biological Characteristics, June 1984, Astoria, Ore.)

disrupt the stability of the watersheds. Such disruption increases erosion and thus the amount of sediments carried down streams and rivers, which in turn increases the rate at which sediments can accumulate in the estuaries.

CHAPTER 6

THE SEA

Prior to civilization and its technology, pelagic fauna congregated around large driftwood and other floating vegetation; floating carcasses of such dead marine animals as whales, seals, sea lions, and sea turtles; and such huge living organisms as whale sharks. Today, however, much of the floating material in the world's oceans, in addition to seaweed and driftwood, is human garbage dominated by styrofoams, petrochemical products, and fishing gear debris.

DRIFTWOOD IN THE OPEN OCEAN

According to a recent survey, most floating debris in the eastern Pacific is plant material, primarily trees, branches, or other tree parts, including palm, banana, and mangrove trees (many of which are a result of logging) as well as bamboo and other types of canes. Another significant portion of the debris is wooden objects originating from human activities: pallets, planks, plywood, boats and parts thereof, rafts, and cable drums. The rest is an assortment of human garbage.[1] In the eastern tropical Atlantic, large driftwood is of prime importance.[2]

In the north Pacific, drifting trees that escape the inshore oscillations of the tidal currents enter the open ocean (Figure 6.1), where they may eventually contact the westward transport of the north Pacific gyre, a great circular vortex.

FIGURE 6.1
Log floating in the open waters off the San Juan Islands off the Washington coast.
(Photograph by Chris Maser.)

Once in the north Pacific vortex, large drifting trees can remain afloat for long periods and great distances, coming ashore in such far-off places as the Hawaiian Islands, where in fact trees from the Pacific Northwest accounted for most of the large driftwood on the beaches. Other drifted trees on the beaches of the Hawaiian Islands are native to the Philippines, Japan, or Malaya.

Anthropological records show that the beached Douglas-fir and coast redwood were even integrated into the customs and rituals of oceanic cultures. Ancient Hawaiian civilizations prized the huge Douglas-firs and western redcedars washed up on their shores because local chiefs preferred them for construction of their large double canoes, symbols of wealth and power.

Further south, the coast of the eastern Pacific, particularly along Central and South America, receives a great deal of precipitation during the monsoon season, which begins in Peru and sequentially heads northward to Columbia and Panama, where it occurs about two months later than in Peru. Even the relatively dry Mexican coast can experience a goodly amount of rain during any given month of the rainy season, usually in the form of short tropical storms. Thus, the floods bearing driftwood to the sea occur at different times in different places.

Although it rains throughout most of the area, there are three places of high annual precipitation that affect the nearshore of the eastern Pacific along Central and South America. These are, in order of importance, the coasts of Colombia and northern Ecuador, the coast of Costa Rica and Panama, and the coast of the Guatemala–Mexico highland. Although the most important rivers in the region discharge into the Caribbean or the Atlantic, many smaller rivers discharge into the eastern Pacific, and therefore, river runoff is important during the rainy season.[3–5]

Rainy seasons vary widely among years, however, due in part to the pattern and intensity of El Niños. An El Niño is a regional or global oceanic–atmospheric disturbance whose manifestations range from increased sea surface temperature in the tropical east Pacific to aberrant rainfall patterns. For example, as one of the earliest proofs of very strong El Niños, severe flooding occurred in the Moche Valley on the north coast of Peru about 500 B.C. Another cataclysmic flood, the Chimu Flood, occurred during the Chimu dynasty, within a century of the year 1100 A.D.[6, 7]

Other evidence of El Niños includes: (1) significant differences in travel times of sailing vessels between ports along the Peruvian coast (northward travel took longer than normal and southward travel time was greatly reduced due to southward coastal currents and/or winds); (2) data from the ship logs of pirates, privateers, and explorers, indicating unusual sea and weather conditions, unusual sea and air temperature, or displaced continental vegetation floating on the sea; (3) penetration of abnormally warm waters farther south than usual along the coast of Peru, the summer and/or autumn in the southern hemisphere; (4) thunderstorms, heavy rainfall, and/or flooding; and (5) destruction of buildings and sometimes whole cities in the coastal zone by flooding.[8]

Prior to deforestation, the amount of rainfall and the consequent magnitude of flooding determined how much wood got out to sea, where winds and currents decided its distribution. An El Niño event can have a dramatic effect on both the amount of driftwood discharged from a pristine river in a given year and the distance and direction it might travel once at sea before it becomes water-logged and sinks.

The logs of travel times from early sailing ships can provide some insight into how driftwood can occasionally cover vast distances in short periods. For example, favorable winds occurred in early 1531, allowing Captain Pizarro to make a rapid 13-day trip from Panama to the Bay of San Mateo, a trip that had previously taken him two years to complete (beginning in November 1524) because of adverse conditions created by a strong El Niño from 1525 to early 1526.[9] In another such trip, Padre Geronimo Ruiz Portillo and six companions journeyed in 1568 from the Port of Panama to Lima, Peru, in 26 days, a trip that usually took six months.[10]

In the tropics in general, however, some 70 percent of the world's continental water and 74 percent of the world's sediments are discharged from rivers. For example, the world's three largest rivers empty into the west Atlantic; in order of decreasing size, they are the Amazon, the African Congo, and the Orinoco, which flows north of the Amazon.[2, 11]

It therefore appears likely that most driftwood entering tropical oceans comes from local forests and is transported to the sea by local rivers. Several sources of river-borne driftwood are possible. The most pristine is driftwood discharged from virgin tropical forests, where flood waters undermine trees along river banks, causing them to topple into the current and be washed out to sea, or debris floods and avalanches on small streams disgorging large amounts of wood into a large river and then out to sea during a flood. Other possibilities are human exploitation of jungles for forest products, the escape of river-transported timber, the clearing of riverbanks, or the diverting of river flow patterns. Deforestation of river basins, as has occurred within historical times for many large Asian rivers, adds much driftwood to the ocean for a time and may also change the rivers' hydrological regimes, which, in conjunction with the loss of forests, may result in the disappearance of driftwood altogether.

Short rivers that drain primary rain forests with relatively minor alluvial plains and enclosed estuaries, such as the Fly, Sepik, and Tami rivers of Papua New Guinea and the Paz, Negro, and Mira rivers of western Central America, are important. In fact, swift, short rivers may have an importance in the discharge of suspended materials that belie their modest size, for as recent data suggest, the Fly River, which despite its short length is tenth in the volume of water in the world delivered to the sea.[12] When such rivers are coupled with limited estuaries that have high rates of flushing, great amounts of wood are annually discharged into the sea.

Large rivers with cultivated, mainly deforested lower floodplains and well-developed estuaries with low rates of flushing may be of reduced importance as a source of driftwood. To date, this may not be true for the Amazon. Although active deforestation is taking place, it has not yet reached a point characteristic of some large Asian rivers, such as the Brahmaputra, which flows 1800 miles from southwestern Tibet to form a delta with the Ganges in the Bay of Bengal. These rivers carry a high silt burden due in part to deforestation in their lower basins.

Because most driftwood discharged from tropical rivers occurs during monsoon floods, its distribution once in the ocean varies in time and space. Despite its distributional variability, ocean currents keep most of it in the coastal zone close to its source, even after a full year at sea, during which time much of it becomes water-logged and sinks.

In the western Pacific, for example, driftwood is distributed mainly in the coastal areas, with a zone of lower density between 10 and 30 degrees north. Driftwood is still relatively abundant in the tropical areas, such as the waters around the Philippines, Indonesia, Papua New Guinea, and the Solomon Islands, as well as along the coast of central Mexico. In the eastern tropical Atlantic, on the other hand, a major driftwood concentration is located at 2 degrees north and 15 degrees west in an area of convergence, but it is unknown if the driftwood originates from African rivers or the Amazon.

Unless the influx of new drifting trees is allowed to remain at historical rates, large driftwood will continue to disappear from the world's oceans and will someday be extinct. Although it may not seem so, extinction of large driftwood may be imminent, due to the continuing deforestation of both tropical and northern temperate forests. Already the total volume of driftwood is lower and the individual size classes of drifting wood are smaller due the industrial philosophy of total utilization of wood fiber. In fact, the mighty Colorado River, which used to carry much driftwood into the Gulf of California, now flows through so many dams and has lost so much riparian forest that driftwood no longer rides its current. Even if it did, it would not reach the ocean because the Colorado is sucked dry before it gets there.

Coastal deforestation in recent decades, particularly of mangrove forests, ranges from 50 percent in the Philippines to 20 percent in Malaysia.[13] Although mangrove forests may not be a major source of large driftwood, given the generally high density of their wood, which causes it to sink, they indicate that deforestation close to coastal areas is proceeding at a pace greater than inland.

Iceland, for example, had its ancient forests already "improvidently exhausted" by the mid-1800s, according to geologist Sir Charles Lyell:

> ...although the Icelander can obtain no timber from the land, he is supplied with it abundantly by the ocean. An immense quantity of thick trunks of pines, firs, and other trees, are thrown upon the northern coast of the island, especially upon North Cape and Cape Langaness, and are then carried by waves along these two promontories to other parts of the coast, so as to afford sufficiency of wood for fuel and for constructing boats. The timber is also carried to the shores of Labrador [the mainland part of Newfoundland, Canada] and Greenland; and Crantz assures us that the masses of floating wood thrown by the waves upon the island of John de Mayen often equal the whole of that island in extent. [John de Mayen is now called Jan Mayen and is a fairly large island of Norwegian ownership lying north northeast of Iceland and east of Greenland.]
>
> In a similar manner the bays of Spitzbergen [a Norwegian archipelago in the Arctic Ocean between Greenland and Franz Josef Land] are filled with drift-wood, which accumulates also upon those parts of the coast of Siberia that are exposed to the east, consisting of larch trees, pines, Siberian cedars, firs, and Pernambuco and Campeachy [sic] woods [which means trees from Pernambuco, a state in northeastern Brazil, and Campeche, a state in southeastern Mexico on the western part of the Yucatán peninsula]. These trunks appear to have been swept away by the great rivers of Asia and America. Some of them are brought from the Gulf of Mexico, by the Bahama stream, while others are hurried forward by the current which, to the north of Siberia, constantly sets in from east to west. Some of these trees have been deprived of their bark by friction, but are in such a state of preservation as to form excellent building timber. Parts of the branches and almost all the roots remain fixed to the pines which have been drifted into the North Sea, into latitudes too cold for the growth of such timber....[14]

As previously indicated, the fate of driftwood discharged from rivers and estuaries into the open ocean is determined by winds and currents. Through this mechanism, aboriginal peoples of the far north have been supplied with driftwood from the shores of rivers and the sea far above the arctic tree line for fuel and building materials for summer shelters.

Because these currents vary along the Oregon–Washington coast in direction and intensity with season, the direction in which wood drifts from a particular estuary varies greatly.

DRIFTWOOD, CURRENTS, AND BEACHES

The broad California surface current flows south along the coast all year, at about two tenths of a mile per hour, but it increases in speed during summer when the winds at the surface of the water are from the north-northwest (Figure 6.2). During winter, a second coastal flow, called the Davidson current, forms and flows northward over the Continental Shelf inshore of the California current, pushing the California current offshore. Average velocity of the Davidson current (about half a mile per hour) results from strong northerly air circulation in the atmosphere.

Currents inshore of the California current depend on seasonal changes in prevailing winds, which drive the flows at the water's surface and transport driftwood. In response to changes in wind direction, inshore currents along the Oregon–Washington coasts vary in speed from 2 to 12 inches per second and in direction from predominantly south in winter to predominantly north in summer. Each phase of the current's direction lasts from four to five months and is interrupted with short, transitional periods.

Coastal winds shift seasonally in their north-to-south direction with transitional periods of weaker winds. Strong winds, blowing predominantly out of the southwest from October through April, increase to gale force during winter storms and move a shallow layer of surface water northward along the coast and toward shore. These winds often beach driftwood, including large trees discharged out of the mouths of estuaries. Stumps and stems of redwood trees and other driftwood from northern California become stranded on Oregon beaches when tides and winds combine to cause unusually high water along the coast.

In summer, offshore movement of water from the nearshore zone is often balanced by the upwelling of cold deep water in a zone five to ten miles offshore. Complex shoreward movements of water associated with these fronts of upwelling may entrain driftwood and retain it in the coastal region.

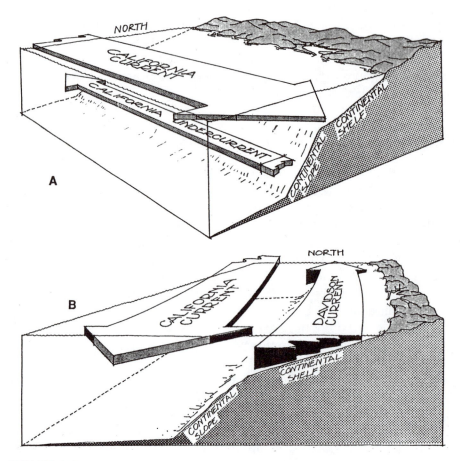

FIGURE 6.2

Nearshore currents transport driftwood northward along Oregon coastal beaches. (A) Summer circulation off Oregon. The California current, a broad, shallow surface current, drifts slowly southward over the Continental Shelf and Slope during the summer. The California undercurrent is a narrow, fast-moving current flowing northward at depths greater than 650 feet over the Continental Slope. (B) Winter circulation off Oregon. The fast-moving, relatively narrow Davidson current flows northward at all depths over the Continental Shelf. The California current, flowing slowly southward on the surface, is pushed offshore by the Davidson current. (Reprinted with permission from Parmenter, T. and R. Bailey. 1985. *Oregon Ocean Book,* Oregon State University, Corvallis.)

Driftwood returned to the coastal portion of the ecosystem by currents and winds may be driven ashore by the southwesterly winter storms, where it is repeatedly beached and refloated by high tides. Deposits of driftwood have altered the evolution of the coastal shoreline of the Pacific Northwest during this century by temporarily protecting isolated cliffs and building beaches where they otherwise might not have occurred. This process also alters the species composition and patterns of plant succession on the rocky beaches.

Driftwood deposits along the Pacific Northwest coast differ in kind and quantity from those along other coasts and may cover many acres of beach to a depth exceeding ten feet. In the mid to late twentieth century, the total accumulation of driftwood between Cape Mendocino in California and Puget Sound in Washington was estimated to be several billion board feet.

Drifted trees play an important role in the cycles of erosion and deposition of sandy beaches and foredunes along the Pacific Northwest coast, where they are driven ashore by wind and waves to form huge piles until removed by extremely high water. Driftwood acts as a barrier to wind-transported sand, forming a nucleus for sand accumulation (Figures 6.3 and 6.4). Drifted trees are thus found deep in the bases of large sand dunes (Figure 6.5) and sand spits until another cycle of waves erodes the sand, returning the driftwood to the sea to be deposited elsewhere.

Drifted trees thus act as passive stabilizing objects along the fronts of coastal dunes. Accumulations of drifted trees protect isolated cliffs from wave action and allow beaches to grow where they would not otherwise be. Those drifted trees beached beyond the reach of waves among the foredunes are not subject to continued deterioration by marine wood-borers, but they, like trees recumbent in moist, terrestrial environments, are rapidly attacked by wood-decaying fungi and decompose in a similar manner. Because large driftwood offers shelter and shade, colonizing plants often begin growing in the moist sand along its protecting edge.

DRIFTWOOD BATTERING ROCKY SHORES

Rocky shores are another site of rapid deterioration of large driftwood into fragments small enough to enter the detritus-based energy

FIGURE 6.3
Driftwood on sandy beaches forms nucleus for temporary accumulation of sand.
(Photograph by Lewis Nelson.)

system. Driftwood beached on rocky shores (Figure 6.6) undergoes a different set of processes than does that beached on sandy shores.

Storm waves pile large quantities of driftwood against the sides of rocky points and headlands. Even massive, drifted trees in this situation are scarred from having been severely battered as waves drive them against the rocks at high tide, splintering the wood so that shores abound in freshly broken fragments. Much of the partly water-logged, splintered,

FIGURE 6.4
Sand accumulating around base of driftwood. (Photograph by Chris Maser.)

and shattered material sinks, only to undergo further grinding into coarse debris, some of which lodges amongst and below the rocks while most is washed into deeper water by pounding waves.

Driftwood plays an important ecological role in the structure of the biological communities of the rocky, intertidal shores, where rock-dwelling plants and animals compete for space in which to attach themselves. In the absence of other influences, the more successful or dominant competitors gradually exclude the weaker or subordinate ones, occupying all available rock surfaces, thus decreasing community diversity.

Battering by drifted trees during the high winter tides helps maintain the diversity of intertidal communities on the rocky shores. The trees strike the rocks with sufficient force to kill the plants and animals attached there, creating unoccupied space.

The role driftwood plays in the processes of biotic communities on rocky shores is proportional to its abundance and size. How often an area is battered by large driftwood depends on its location. Any given spot in

FIGURE 6.5
Three large pieces of driftwood recently exposed by a violent winter storm after having been buried for many years deep within the base of an oceanfront sand dune. (Photograph by Chris Maser.)

the intertidal shore of San Juan Island, Washington, for example, has a 5 to 30 percent chance of being struck by drifting trees at high tide over a three-year period. Nothing is known, however, about the long-term effect of the current absence of large drifting trees from this portion of the ecosystem.

DRIFTWOOD AS HABITAT AND FOOD

Trees drifting at sea are usually heavily populated with plants and animals, including both pelagic and nearshore species, acquired during the tree's transit to the open ocean. Once in the open ocean, other organisms use them.

The ocean strider (Figure 6.7), the only insect known to successfully invade the open ocean, alternately floats and skates on the surface

FIGURE 6.6
Driftwood battering rocky shores is important. It kills organisms already inhabiting the rock and creates new habitats for attachment of intertidal organisms, as well as providing small splinters of wood to the coastal food chain. (Photograph by Lewis Nelson.)

FIGURE 6.7
Ocean strider with prey skating on the open sea. (Photograph ©Lanna Cheng, Scripps Institution of Oceanography, University of California, San Diego.)

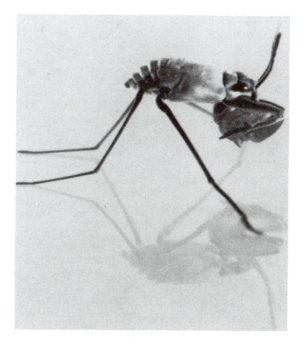

of the sea and attaches its eggs to floating driftwood. Lacking wings, these skaters of the sea live exclusively on the ocean's surface, either drifting passively with the winds and currents or skating actively on the waves.

In addition to the communities of organisms actually attached to or secreted in the driftwood itself, communities or other organisms form around it. More than 100 species of invertebrates and some 130 species of fishes are known to congregate on and around different types of floating objects. With respect to these open-ocean communities, wind-induced water movements in the ocean's upper layer are among the most important factors leading to the aggregation and survival of organisms near floating objects.

The rapidity with which fishes aggregate around driftwood has to do with the Langmuir currents, which are parallel pairs of counter-rotating convection currents driven by surface winds (Figure 6.8). Langmuir convection currents sweep driftwood, organic detritus, and plankton into the long, parallel windrows (often called slicks) of floating debris or lines of foam.

Various hypotheses have been advanced to explain the formation

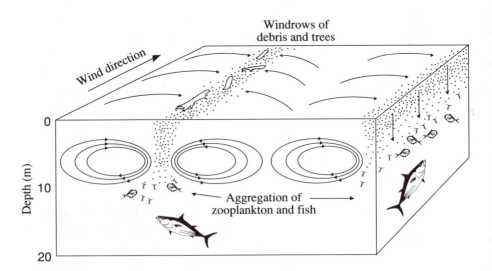

FIGURE 6.8
Patterns of circulation of wind-driven Langmuir currents, which result in windrows (also called slicks) of driftwood and foam.

of communities associated with floating objects, particularly fishes. Shade, refuge from predatory fishes, an abundance of food, or protection from waves may all be factors. It has also been suggested that shadows cast by floating trees make zooplankton more visible to predators or that pelagic fishes simply seek shade. Then again, drifting trees may serve as sites for egg attachment, sources of algal and invertebrate food for pelagic fishes, or places where they can clean external parasites off themselves.

In addition, it has been hypothesized that small fishes are initially attracted to a floating tree as a point of reference, swimming with it while feeding on the small planktonic organisms in its vicinity. Along with the planktivores come juveniles of carnivorous species. With time, larger obligate predators arrive and eat the accumulated prey. Thus, within three to five weeks after a tree initially arrives in the open ocean, having collected plankton and attracted first small fishes and then larger, predatory fishes, such as dorado and tunas, the combined weight of its associated tunas alone may reach as much as 100 tons.

The prey responds by using the floating tree as protection. The predators rapidly deplete the available prey, albeit only temporarily. In their search for food, the predators move away from the tree only to return and use it as a point of reference.

In the eastern Pacific, the primary species/groups aggregating around driftwood are small yellowfin tuna, skipjack tuna, dorado, and sharks (probably silky sharks). Dorado, small yellowfin, and skipjack are also the main species associated with driftwood in other oceans, which indicates its importance in their life histories.

Tunas even time their migration to the Continental Shelf for spawning to coincide with the onset of monsoon rains. In turn, the resulting floods, carrying new driftwood to the sea, arrive as the young tunas are hatching from their eggs. In the eastern Pacific, for example, it is very likely that the association of juvenile yellowfins with large driftwood is important in determining recruitment success. Skipjack, on the other hand, which spawn only infrequently in the eastern Pacific, also have a strong tendency to associate with large driftwood. In addition, the association of some species with old driftwood with barnacles and algae attached may indicate a degree of habitat specificity in the association.

Large driftwood in general plays an important role in the life cycles of many eastern Pacific animal species. It attracts the epipelagic fauna in the region, those in the upper surface layer of the open ocean, and in some cases probably most of the available species. There is a greater diversity and abundance of fauna associated with such floating objects as large driftwood and with the water patterns in which they are found, such as convergence zones and Langmuir currents, than in the entire rest of the open ocean.

Large driftwood attracts an aggregation of species that would otherwise be dispersed throughout the pelagic environment, aggregations similar to those around islands and seamounts. They are analogous to the faunal aggregations around reefs and seagrasses in coastal settings.

Epipelagic fauna associate with large driftwood for various lengths of time and/or at different stages of their life cycles. The abundance of juveniles of many species indicates the importance of driftwood in the recruitment of new individuals into populations. Further, the species composition and relative frequency of occurrence within aggregations associated with driftwood in the eastern Pacific probably depends on the proximity of refugia, such as islands and seamounts.

Faunal aggregations in different areas often vary in species composition, even when the areas in question are immediately adjacent to each other. In such places, the difference in species composition is probably related to the juxtaposition of circulation patterns and water masses, such as the California current being next to a warm-water tropical area.

In fact, such floating objects as large driftwood behave essentially as free-drifting islands or reefs. The ecological role played by these free-floating islands[1] is probably different for different species because the assemblage is dynamic, changing in space and time, albeit with some degree of organization. The faunal aggregation may show different developmental stages (for example, changes in species composition) as the driftwood is carried by the current across hundreds of miles over weeks and months.

It thus seems that driftwood can be adapted by many species for shelter or food, but for most species driftwood is the vehicle that brings them into rich feeding grounds. Driftwood, originating at river mouths, travels through biologically rich coastal areas and tends to accumulate in rich parts of the ocean. In this way driftwood can be used as stepping stones in the movements of tunas, dorado, and other large fishes because it serves as a physical link between major features in the circulation of ocean currents and the fishes' migration paths and the location of a region's islands and seamounts.

Assemblages of pelagic fauna around seamounts and islands separated by expanses of deep water are analogous to the associations of birds and mammals on oceanic islands. In the biogeographic sense, driftwood acts as free-floating islands, carrying subsets of epipelagic fauna from one biologically rich area to another, a process that can serve to further the ecological study of pelagic faunal assemblages associated with large driftwood in the open oceans of the world.

The analogy of driftwood to free-floating islands is especially apropos to the floating wood islands that were once created in rivers, such as the Mississippi:

> One of the most interesting features in the great rivers of this part of America [Louisiana] is the frequent accumulation of what are termed "rafts," or masses of floating trees, which have been arrested in their progress by snags, islands, shoals, or other obstructions, and made to accumulate, so as to form natural bridges, reaching entirely across the stream. One of the largest of these was called the raft of the Atchafalaya, an arm of the Mississippi, which was certainly at some former time the channel of the Red River, when the latter found its way to the Gulf of Mexico by a separate course. The Atchafalaya being in a direct line with the general direction of the Mississippi, catches a large portion of the timber annually brought down from the north [which Lyell later says is a "prodigious quantity"]; and the drift-trees collected in about thirty-eight years previous to 1816 formed a continuous raft, no less than ten miles in length, 220 yards wide, and eight feet deep. The whole rose and fell with the water, yet was covered with green bushes and trees, and its surface enlivened in autumn by a variety of beautiful flowers. It went on increasing till about 1835, when some of the trees upon it had grown to the height of about sixty feet. Steps were then taken by the State of Louisiana to clear away the whole raft, and open the navigation, which was effected, not without great labor, in the space of four years.[14]

Once formed, these wood islands often reached the open sea:

> Within the tropics…there are floating isles of matted trees, which are often borne along through considerable spaces. These are sometimes seen sailing at the distance of fifty or one hundred miles from the mouth of the Ganges [which empties into

the Bay of Bengal], with living trees standing erect upon them. The Amazon, the Congo, and the Orinoco, also produce these verdant rafts, which are formed in the manner already described when speaking of the great raft of the Atchafalaya, an arm of the Mississippi, where a natural bridge...has existed for more than forty years,...rising and sinking with the water which flows beneath it. That this enormous mass will one day break up and send down a multitude of floating islands to the gulf of Mexico, is the hope and well-founded expectation of the inhabitants of Louisiana....

In a memoir lately published, a naval officer informs us, that as he returned from China by the eastern passage, he fell in, among the Moluccas [formerly the Spice Islands, a group of Indonesian islands lying between Sulawesi and New Guinea], with several small floating islands of this kind, covered with mangrove-trees interwoven with underwood. The trees and shrubs retained their verdure, receiving nourishment from a stratum of soil which formed a white beach round the margin of each raft, where it was exposed to the washing of the waves and the rays of the sun. The occurrence of soil in such situations may easily be explained, for all the natural bridges of timber which occasionally connect the islands of the Ganges, Mississippi, and other rivers, with their banks, are exposed to floods of water, densely charged with sediment.

Captain W. H. Smyth informs me, that when cruizing [sic] in the Cornwallis amidst the Philippine Islands, he has more than once seen, after those dreadful hurricanes called typhoons, floating islands of wood, with trees growing upon them, and that ships have sometimes been in imminent peril, in consequence of mistaking them for terra-firma, when, in fact, they were in rapid motion.[14]

The fact that free-floating islands of wood exist or have existed within the world's major rivers and in the convergence zones of the open ocean is well documented all over the world. The ecological role of these islands, which can be a single large tree or an aggregation of large trees, is just now being recognized and studied by ecologists and resource managers. Sailors and fishers, however, have for centuries known about the uniqueness of these floating islands for food fishes.

DRIFTWOOD AND TUNA

When purse-seine nets were adopted by the tuna fishery, it gave rise to the three primary methods of purse-seining for tuna in use today:

1. *School fishing.* Tunas are detected from signs on the water's surface that are visible from a vessel or helicopter. These signs include a school feeding or swimming rapidly close to the top, which disturbs the surface. A school may also have its presence betrayed by a flock of birds, or the fish may be seen jumping.

   Types of schools are differentiated by the effect of their behavior on the surface of the water. A "breezer," for example, is a school that affects the water's surface as if it is being blown by a breeze. A "boiler," on the other hand, makes the water look as if it is boiling.

   Although fishers call this type of fishing "school fishing" or "fishing on schools," the terminology is misleading because schools are the target of several modes of fishing. School fishing should be understood as any kind of fishing not associated either with floating objects, such as driftwood, or with dolphins.

2. *Dolphin fishing.* "Dolphin fishing" takes advantage of the association of large yellowfin tunas with herds of dolphins. By detecting the easily visible, surface-swimming dolphins, chasing them, and maneuvering them into the net, fishers capture the tunas as well. They are so closely associated with the dolphins that a school stays with the dolphins throughout the chase and encirclement.

   Once caught, the tunas are retained, while the dolphins are supposedly released. Sometimes, however, not all the dolphins are released, and since dolphins are air-breathing mammals, those retained in the nets drown.

3. *Log fishing.* In log fishing, a fisher searches for floating objects, such as large driftwood, under which a school of tunas is gathered. A net is then set around the object and the fish are captured.

Japanese and American fishers have long known about such aggregations around large driftwood (Figure 6.9). With their success rate well over four to one in favor of netting schools of tunas and with tuna such a valuable commodity, they routinely seek large floating driftwood to pursue their trade. In fact, the importance of this knowledge is illustrated by the success of the purse-seine tuna fishery business. In the western and central tropical Pacific, it evolved from almost nothing in the mid-1970s

FIGURE 6.9
Drifting wood in the open ocean serves as a focal point for zooplankton and small fish. Large predators, such as billfish and tuna, often congregate around such floating debris.

to the world's largest in both total catches and numbers of boats deployed within a decade or so after discovering that tuna schools associate with such things as large floating driftwood.[15]

But how do tunas find driftwood in the first place? It is believed that the various species are visual predators that forage only during hours of daylight and twilight, and possibly under the light of a full moon. When not feeding, they tag along with such floating objects as large driftwood in order to find and remain in water masses rich with food. Although it is not known how tuna detect surface objects like driftwood, whether through visual, olfactory, or sonic detection, the fact that this behavior occurs in all oceans and in many different species of fishes shows how successful an adaptation it is.

If tunas did not accompany floating driftwood but instead swam randomly to find food, it is quite possible they would find themselves in waters poorer in food then the ones they had left. In many cases the transition from rich to poorer areas is abrupt, making it quite possible that tunas will enter one of these poorer areas during random swimming. This is particularly true in those parts of oceans where the most productive areas are narrow, often along coastlines near the mouths of big rivers or relatively near shore.

Current patterns keep most driftwood in coastal zones near its sources, even after a full year at sea, during which time much of it becomes water-logged and sinks. Driftwood that does float offshore in the eastern Pacific, for example, ends up in a rich frontal zone rather than in the less productive central gyres to the north and south of that zone. Driftwood thus acts as a retention mechanism, keeping tunas for a time in the rich coastal areas before leading them westward through the most productive feeding grounds of the eastern Pacific.

Knowing that tunas associate with large driftwood, fishers search for it. Once it has been located, either visually or by "bird radar" (since birds are frequently observed near floating driftwood, at least in the eastern tropical Atlantic), the boat approaches it to evaluate the school's characteristics, such as size and species, either by sight or by sonar. Some skippers, however, prefer not to approach driftwood directly with their vessels, so they launch instead a small boat with a spotter to evaluate the possible catch.

If it is decided that a set should be made, the vessel is moved to a certain distance from the driftwood, which will be in the center of the area's circumference as it is surrounded by the net. The auxiliary skiff, which carries the end of the net, is then lowered and begins circling the driftwood. Once the circle is complete, the purse line is drawn and the net gathered until the catch can be scooped out.

Before recovering the fishing gear, however, an auxiliary launch goes to the driftwood and attaches a line to it, pulling it carefully out of the circle. It is then left near the tuna vessel until the result of the catch is evaluated. Depending on the result, the skipper will make one of the following decisions: (1) abandon the driftwood and continue searching for new schools; (2) place a streamer or a radio range beacon on it, with the idea of returning at a later time, and then continue searching; or (3) remain next to it so it cannot be used by other vessels and proceed catching the tunas associated with it whenever it is felt the time is right, generally dawn of each day. Although the Japanese in the western and central Pacific may also set nets daily around the same piece of driftwood, they may at other times set their nets repeatedly but allow a few days to elapse in between sets.

If the driftwood is found near the continental shelf and the currents are moving toward the shelf, a long cable is placed from the tuna boat to the driftwood and another to an auxiliary launch or skiff. This launch functions as a towboat for the tuna vessel, which stops its motors. The movement of the driftwood by the current is thus counteracted, preventing it from reaching the continental shelf. This prevents damage to the net in successive fishing operations. This is important for the Spanish tuna fishery in the eastern tropical Atlantic, whose nets reach to a depth of 650 feet.

There are variations on this general theme of "log fishing" for tunas. Sometimes the tuna boat itself, as it remains adrift at night, attracts tunas beneath it. Tunas have also aggregated under anchored auxiliary skiffs. In the beginning of the Spanish tuna fishery in the eastern tropical Atlantic, some companies operated with the help of *maciceros*, boats from which bait was thrown to keep tunas in the area.

Toward the end of 1990, Spanish tuna fishers in the eastern tropical Atlantic began generalized log fishing by supplementing large driftwood with artificial logs. These artificial logs are square bamboo rafts about five feet in diameter and approximately a foot in height. The bamboo pieces themselves are about eight inches in diameter. Each bamboo raft is equipped with a black net cloth covering the top and a buoy with a radio range beacon.

These beacons can either emit signals continuously or be sleepers, that is, begin emission only in response to a signal from the tuna boat. Each boat's signal, however, is different and its frequency kept secret.

In addition to the already completed rafts each tuna boat carries as it leaves Abidjan, a seaport on the Gulf of Guinea and the capital of the Ivory Coast, at least three auxiliary boats accompany it. Each auxiliary boat's purpose is twofold: (1) make bamboo rafts while on the high seas and (2) continually monitor the beacons of rafts already deployed and alert the tuna boat if tunas are associated with a particular raft. The tuna boat, in turn, moves from one raft to another during the night to increase the efficiency of its efforts.

In general, the species composition of tunas associated with driftwood and bamboo rafts is quite stable from year to year and place to place. Skipjack is the dominant species, followed by yellowfin and bigeye. A factor of major importance with respect to the artificial log or bamboo raft technology is that the fishing zone for skipjack has been extended to the offshore area south of the equator and west of ten degrees west, an area where this species was unavailable to any fishery until 1990. Consequently, a heretofore hidden fraction of the skipjack stock is now suddenly available for exploitation through the new bamboo raft technology.

Bamboo rafts deployed in convenient areas concentrate tunas with characteristics similar to those associated with large driftwood: large, easily caught, multispecies schools. The major difference between driftwood and bamboo raft fishing is that fishing around driftwood is seasonal and geographically restricted, whereas bamboo rafts can be efficiently and effectively deployed in a greater area over a longer fishing season.

What do such technological advances as bamboo rafts fitted with radio transmitters mean for the future of the tuna fishery in the eastern tropical Atlantic, or any other ocean? The most obvious consequence is that once again society has the potential for the overfishing and possible extinction of yet other species. It is primarily young tunas that are caught under driftwood and bamboo rafts, which most likely has an effect on the recruitment of new individuals into the breeding population. Thus, where driftwood by its very nature once helped to keep the fishing pressure within some limits, technology has effectively removed those limits but has done nothing to increase the social wisdom with which tunas are now exploited.

BORERS OF THE DEEP

Although the sources, forms, and routes that supply organic material to the energy-poor deep sea are little known, it is known that drifting trees in nearshore currents, usually inhabited by northern gribbles and native shipworms, become water-logged and sink over the continental shelves or in the deep sea, where they represent terrestrially fixed carbon. In the Pacific Northwest and other oceans, deepsea wood-boring bivalves live on the ocean floor, where they depend on sunken driftwood for survival. These borers of the deep are a separate subfamily (Xylophaginae) from the shallow-water shipworms described earlier. Deep-sea wood-borers are temperature-sensitive animals that live at depths greater than 6000 feet in the tropics, but can exist in shallower waters at northern latitudes. Although they probably exist in all oceans, they have been studied in only a few areas.

There are at least three species of deep-sea wood-borers in the genus *Xylophaga* of the family Pholadidae. Deep-sea wood-borers quickly invade wood and grow rapidly. In the Atlantic Ocean of Long Island at a depth of about 6000 feet, wood can become infested with a dense population of two species within 104 days (Figure 6.10).

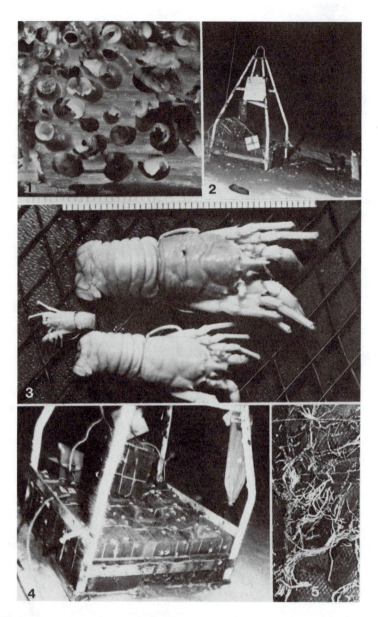

FIGURE 6.10

Colonization of sunken driftwood by deep-sea wood-borers more than 6000 feet beneath the surface. (1) Surface of wood removed to show concentration of borers (Xylophaga) and polychaete worms (scale in millimeters). (2) "Wood island" composed of 12 blocks of wood in metal frame on ocean bottom. Note two wooden panels on right of island. Staining on bottom around island is probably accumula-

Such dense populations convert wood into fecal pellets, which settle to the sediment surface and attract other bottom-dwelling animals. The conversion of wood into a readily available source of detritus supports the development of a complex local community of bottom-dwelling organisms.

Deep-sea wood-borers have a high reproductive rate, grow rapidly, mature early, and have high-density populations, all of which are characteristics suitable for effectively using wood, the supply of which is erratic, unpredictable, and transient in the deep ocean. Their piles of fecal pellets, which are finely ground wood fragments, may attract more than 40 species of other deep-sea invertebrates (Figure 6.10, panel 2). Enrichment of the bottom, a result of disintegrating wood and the accumulating fecal pellets, thus contributes to the development of a rich fauna.

Because adults die as wood is consumed and disintegrates, dispersal must be through an abundantly produced, efficient, and long-lived larval stage that actively searches for wood at the interface of bottom sediments and water. Exactly how this wood-dependent deep-sea community finds new wood is unknown, however.

Limited data suggest that spawning in *Xylophaga* is seasonal and that it may coincide with the rainy season in the tropics and winter or spring snowmelt and runoff in northern latitudes. Deep-sea wood-borers seem to be adapted to exploit wood quickly at any time of the year. As such, they are one of a few deep-ocean opportunistic organisms not associated with hydrothermal vents.

Wood in the deep sea supplies energy in an energy-scarce environment and is a source of habitat diversity. It is not surprising, therefore, that a single, water-logged tree on the deep-ocean floor is the focus of intense and abundant life. Although wood is frequently found in dredges from the deep-sea floor and deep-sea trenches, it is most common off mouths of rivers and wooded coastlines.

Although communities of bacteria can use sulfur compounds as energy, and animals can and do thrive around deep-sea hydrothermal vents

---

tion of feces from borers. (3) Deep-sea galatheid crabs fed on whatever was available on wood surface as they grew inside the mesh bag covering wooden panel (scale in millimeters). (4) Close-up of wood island showing large number of crabs on the wooden blocks and thick growth of hydroids on the frame. (5) Close-up of wooden blocks showing growth of serpulid worms (mesh is 10 × 5 mm). (Photographs by Ruth Turner.)

in the ocean's floor, it is commonly believed that the rest of the ocean bottom is sparsely populated with long-lived organisms that have low reproductive potential, slow growth rates, and late sexual maturity. Because these vents, which occur several thousand feet below the surface, have life spans of only decades, the species inhabiting them survive by growing rapidly and producing larvae that disperse widely to colonize new vents. Such vents are thought of as islands of deep-ocean biological diversity.

Similarly, driftwood on the bottom of the abyss functions in much the same way. Each drifted tree on the deep-ocean floor also creates an island of biodiversity and is heavily colonized by wood-boring organisms and their associates. Wood has provided an energy source that has allowed opportunistic organisms to evolve in the stable, slow-changing environment of the deep ocean.

Driftwood, whether in streams and rivers, estuaries, floating islands on the ocean's surface, or wood islands on the bottom of the deep ocean, provides food and habitat on which diverse and productive species assemblages are dependent. Some of the species are dependent solely on wood for their existence; others could find substitute habitats. However, that fact that driftwood moves throughout the world's waterways and that organisms have evolved such a rich diversity of strategies to use it is compelling and fertile ground for further investigation while all of the parts and pathways still exist somewhere in the world.

The role of intact driftwood in the open-ocean food web, whether it is floating on the surface or resting on the bottom, remains unknown, as does the potential role of offshore marine wood-borers in the initial reduction of wood. Nevertheless, continental margins are recognized as the dominant reservoir of organic carbon buried in the marine environment, which is comprised of material derived from both marine and terrestrial sources.

Recent research concerning the biogeochemistry of ocean sediments off the Washington coast suggests that the terrestrial component of the organic carbon-rich sediments along the continental shelf is high. Indeed, the data suggest that particulate input of terrestrial organic carbon worldwide exceeds by a significant factor the burial rate of total organic carbon (both terrestrial and marine) in all ocean sediments. This means that woody materials as well as soil from forested water catchments probably play a significant role as an energy source in the waters near the ocean bottom and in the sediments of the ocean floor.

Most of the particulate organic carbon discharged by rivers is deposited on continental shelves near the point of input, in this case the Columbia River. Ocean currents off the Washington coast act to preferentially trap coarse, lignin-rich, woody plant debris in the shelf environment and allow finer, degraded, lignin-poor, river-derived, soil organic particles to move further off shore. Thus, carbon incorporated in coastal marine sediments changes with distance from shore.

Terrestrial sources account for approximately 60 percent of the total organic carbon in shelf sediments at depths of 150 to 500 feet. In areas where the continental shelf slopes into the deep ocean (1000 to 3000 feet), about 30 percent of the total organic carbon is of terrestrial origin, grading to less than 15 percent in sediments of the deep ocean basin (6000 to 8000 feet).

Although it may take years for large driftwood from the forest to finally reach the sea, the process cannot be summed up in terms of time. It is too limiting for an understanding of the myriad cycling pathways to which driftwood is subjected on its seaward journey.

PART III

FROM THE SEA
TO THE FOREST:
SETTLEMENT

CHAPTER 7

THE SEA AND ESTUARIES

HISTORICAL PERSPECTIVE

George Vancouver, a midshipman with Captain Cook, first saw the Oregon Coast in 1778. He returned 14 years later (1792) in command of two ships sent by the British Admiralty to explore the area claimed by the Spanish. By chance, he met the American sea captain Robert Gray off the Washington coast. Gray told him that he thought he had found the mouth of the River of the West and that he was going back to see if he could navigate it.

Vancouver went on to explore the Strait of Juan de Fuca, sending his sailing master Lieutenant Peter Puget to map the inland waterway that bears his name. Captain Gray did sail his ship, Columbia Rediviva, into the river, which he named after his ship.

In 1805, Lewis and Clark brought additional information about the Columbia River country to the east coast and encouraged American settlement, which stimulated John Jacob Astor, an American, and his five Canadian partners in the Pacific Fur Company to select the Columbia River for the site of their first western trading post. With this early exploration of the Pacific Northwest came the discovery that furs of wild animals, purchased for very little, could be sold in China for immense profit or could be traded for spices, teas, and silks, which brought high prices in both Europe and New England. A competitive trade arose, with ships from England, France, Belgium, Russia, and New England vying for profits. The Americans eventually dominated the trade, partly because the Europeans were preoccupied with Napoleon.

By 1850, the Willamette Valley was becoming well settled, and resident European-Americans had begun in earnest to have an irreversible effect on the land both through agriculture and logging. Reports of the U.S. government and the journals of visitors to the Pacific Northwest document great amounts of large driftwood in the estuaries and on the beaches at the mouths of the rivers. These accounts describe the quantity and size of drifted trees, or snags, which, in the mid-1800s, significantly exceeded present amounts of driftwood in the lower tidally influenced portions of rivers and on beaches. James Swan, who was aboard a seagoing vessel in heavy seas during a storm in 1852 wrote: "The next morning we found ourselves about thirty miles to the westward of the Columbia River, from which a huge volume of water was running, carrying in its course great quantities of drift logs, boards, chips, and saw dust, with which the whole water around us was covered."[16]

The Secretary of the Treasury reported in 1859 that many of the drifted trees in the lower Columbia River were as large as 150 feet long by 13 to 18 feet in circumference. The largest drifted tree found during the survey, conducted in 1858, was 267 feet long.[17] In 1861, James Swan wrote in his journal during a visit to the Northwest Territory: "I noticed...very large trees at the mouth of the [Quillehuyt] River where they had been left on the beach. I measured one of these trees, and found it 250 feet long and eight feet through at the butt. The roots were attached, and spread out some 20 feet....I should estimate that there are two or three millions of feet of lumber in the drift trees at the mouth of this river and vicinity. These have been thrown back from time to time by the breakers on the beach, and have gradually formed a natural levee, behind which the river has forced its way...."[18]

The U.S. Army Corps of Engineers reported that shorelines of estuaries and beaches of the rivers' mouths had often been covered with driftwood in the 1870s (Figure 7.1), as was the mouth of the Coquille River in the 1880s (Figure 7.2).[19, 20]

For several years after coastal areas were settled in the 1850s and 1860s, roads were few and overland travel was impractical, especially in winter. The coastal rivers were under tidal influence for 12 to 40 miles upstream from their mouths. These rivers had low-gradient, deep channels along which commercial boats and log rafts could travel, but slowing currents and storm-wind patterns deposited driftwood in the estuaries.

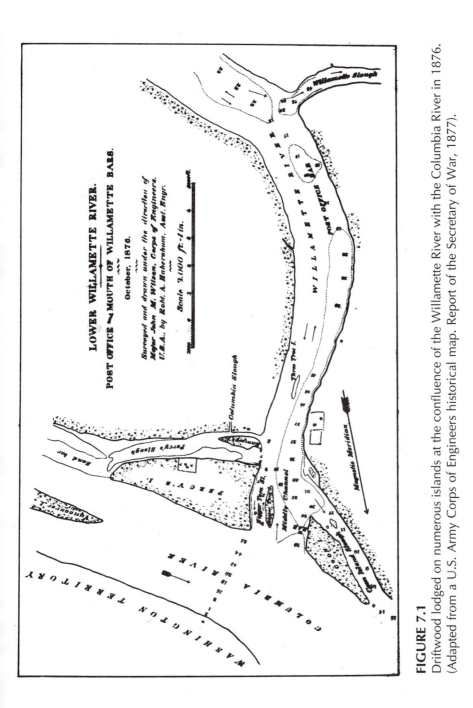

FIGURE 7.1
Driftwood lodged on numerous islands at the confluence of the Willamette River with the Columbia River in 1876. (Adapted from a U.S. Army Corps of Engineers historical map. Report of the Secretary of War, 1877).

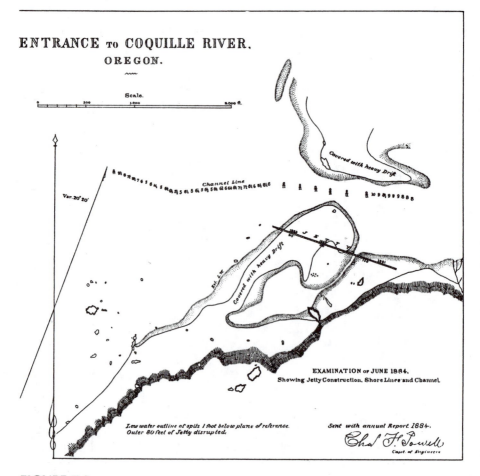

FIGURE 7.2
Driftwood at the mouth of the Coquille River. (Adapted from a U.S. Army Corps of Engineers historical map. Report of the Secretary of War, 1884).

The vast amount of driftwood, including exposed and sunken trees, presented major obstacles for river traffic to negotiate. These deposits of driftwood and sediments in the Columbia River acquired such names as Wood Island and Snag Island.

The U.S. Army Corps of Engineers' responsibility on many rivers during the late 1800s and early 1900s was to improve and maintain the navigability of those rivers deemed economically important. The Corps therefore removed more than 9829 drifted trees from the lower 12 river miles of the Tillamook River, Tillamook County; more than 8591 drifted trees from the lower 22 river miles of the Coos River, Coos County; and more than 6407 drifted trees from the lower 40 river miles of the Coquille River, Coos County, between 1890 and 1917.

During this the 27-year period, the workers pulled a total of 34,827 drifted trees out of the combined 74 river miles in the three rivers, for an average of 336 trees per mile. In addition, more than 86 overhanging trees were cut along the 12 miles of the Tillamook River, 1751 boulders were blasted in the 22 miles of the Coos River, and 81 scow loads of driftwood were removed from the 40 miles of the Coquille River—the latter between 1891 and 1893.[19]

In 1883, the Corps stated of the Coquille River: "From Coquille City to Myrtle Point, a distance of 12 miles, there has been great trouble experienced in keeping the river navigable, owing to snags [drifting trees] and bars. The various forks of the Coquille drain densely-timbered territory, and at every freshet [the increased flow of water during a storm] many trees, stumps, etc., are brought down. Some of these lodge at different points, forming isolated snags, or are grouped together into jams. These snags and jams in turn cause deposits of sand and gravel, forming shoals."[20] The Corps continued to wrestle with the problem of driftwood and the Coquille River's navigability, as evidenced by the 1897 report, which states: "Contractors began active operations on June 18, 1879, at which time sections of the river below Coquille [City] were almost impassable at low tide on account of snags."[19]

The Corps kept tree-removal records in many coastal rivers from approximately 1890 to 1920. Snag boats (Figure 7.3) and dredgers were often used to pull driftwood from the rivers, and blasting large pieces was common, but the Corps' wood-removal operations represented only a portion of the total wood removed from lower rivers. Gill-net fishers formed working teams to remove wood threatening to snag fishing nets. Other local groups or individuals, and later port authorities, worked to maintain channel navigability. Storm-deposited driftwood high on beaches and in the salt marshes was not, however, of immediate concern and was probably left in place.

The driftwood with which these early settlers had to contend was found throughout the estuaries, in channels, on mud flats, and on higher grassy areas. The greatest concentrations of driftwood were along an estuary's shoreline out of reach of the tide, except during storms and very high tides. The most stationary driftwood was at least partly buried in

FIGURE 7.3
The snagboat used to remove driftwood and riverside trees. (Photograph courtesy of the U.S. Library of Congress National Archives.)

marshes and channels. Unburied driftwood in other places was more transient. Although the dynamics of such driftwood are the same, driftwood size is today much smaller.

Driftwood deposited in marshes between the main channel and the shoreline in the lower Nehalem estuary, Tillamook County, has been estimated to have declined since 1939 both in number of pieces (by 50 percent) and in volume (by approximately 60 percent). Stranded wood in the marsh in 1939 probably included escaped logs from lumber companies, which had either been floated downriver or held in rafts.

Timber was commonly floated down coastal rivers to lumber mills prior to development of other practical and inexpensive transportation in the 1960s. Many of these logs, and those stored in rivers adjacent to lumber mills, escaped and sank, adding structural complexity to the estuarine system.

A second human-caused source of wood in water is the large number of pilings and associated wooden structures. In 1986, port authori-

FIGURE 7.4
Log pilings used as pens to hold log rafts waiting to be milled were a major source of human-caused wood in estuaries. (Photograph by Gretchen Bracher.)

ties associated with nine Oregon estuaries estimated that there were at least 15,000 maintained pilings in those systems, with many more thousands of additional wooden structures, such as the old stubs of pilings from pens to hold floating logs (Figure 7.4) and from old channelization projects, old pilings from mills, and the remnants of docks (Figure 7.5).

The open sea was also a source of driftwood in estuaries and on beaches. Winter storms blew driftwood from the sea back into river mouths and onto coastal beaches north of the driftwood's origin. Although some driftwood may have been buried for long periods by river-bottom sediments in estuaries or by sand on beaches, much of it was

FIGURE 7.5
Log pilings for canneries, bridges, and piers were a major source of human-caused wood in estuaries. (Photograph by Lewis Nelson.)

fairly mobile. Still other driftwood was deposited in marshes and along the higher ground bordering estuaries, where it remained until it decomposed.

Today, only 7 of the 17 coastal port authorities in Oregon are directly involved in removing large driftwood in their estuaries. Nevertheless, sources of driftwood have been severely reduced since the 1850s, and opportunities for driftwood to be retained in the lower estuaries have also been reduced through the construction of dikes, the filling in of marshes, and channelization. Pilings create some areas that retain driftwood, but it is likely that driftwood will be cleared away from such sites.

The volume of driftwood on coastal beaches varies among locations and fluctuates between seasons. Winter storms transport wood downriver and into the sea, which then deposits it on beaches. In other seasons, there may be a net loss of wood from beaches back into the sea.

Aerial photographs of river mouths in Oregon show that 43 percent of the existing driftwood volume larger than 16 feet in length was lost between 1970 and 1984. The largest pieces of driftwood measured 52 feet long. The 250-foot drifted tree measured in 1861 in the Washington Territory and the commonly found 150-foot-long drifted trees in the lower Columbia River will not be found today.

Gone are the mature and old-growth stands of coniferous trees, which before the 1800s were the source of the large drifted trees (Figures 7.6 to 7.8). The decline in driftwood on beaches since the early 1970s is probably a result of increased use of fuel wood in wood-burning stoves, which reduces the amount of wood getting into streams. The issuance of Forest Service Free Use Wood Permits by the Northwest Region of the Forest Service (Oregon and Washington) increased sixfold from 1974 to 1982. Further, regulation of forest practices at one time directed that wood be cleaned out of stream channels during logging in the northwest.

FIGURE 7.6
A wild reach of river through virgin old-growth forest. (USDA Forest Service photograph by Jerry F. Franklin.)

FIGURE 7.7
Ever-creeping clearcuts on public lands. (Oregon Department of Fish and Wildlife photograph by Charles Bruce.)

CURRENT LEGAL AND POLICY PERSPECTIVE OF BEACHED DRIFTWOOD

When driftwood does reach a shore or beach, should it be removed or allowed to remain? Beach stability is dependent on the abundance of large pieces of driftwood, which capture land by shielding the shore from wind and wave action. The states of Oregon and Washington, however, have differing views of this question.

Oregon's policy regarding the removal of large driftwood from its beaches reads in part:

> ...to assure continuation of scenic and recreational values for public enjoyment at the ocean shore and to protect marine life and intertidal resources, beach logging, as a general practice, shall be prohibited unless such removal can be shown a significant public benefit.[21]

FIGURE 7.8
The result of industrial forestry on private lands. (USDA Forest Service photograph by Thomas Spies.)

Washington State's policy contrasts markedly with Oregon's. The Department of Natural Resources licenses individuals who retrieve driftwood, either from Puget Sound or from the Columbia River below Grand Coulee Dam. Certain restrictions are imposed on license holders, but the department cooperates by administering the sale of claimed drifted logs by the licensee.

There is little question that a permit holder can simply claim a log floating in Puget Sound. If a log is lying on the beach in front of someone's house, however, ownership is less certain. The Department of Natural Resources says:

Log patrolmen, brand owners, and their agents all have the right to enter peaceably any tideland, marsh, beach, etc. for the purpose of salvaging logs. Likewise beach owners have rights also; where one right ends and the other begins is hard to define. Generally speaking, the removing of logs from the water side by boat does not infringe upon the rights of the beach owner. A beach owner does not have any legitimate claim to a log that washes upon his beach, and to exercise any appreciable claim over this log could result in criminal action.[22]

If the owner of the beach front believes that logs should be left alone to protect the beach from wave erosion, the stage is set for argument. This is not a trivial matter; nine million board feet of drifted logs were appropriated in 1974 by the log patrol. Washington's policy is more a blanket policy than Oregon's because Washington's beaches are considered to be private property, whereas Oregon's are considered to be public property.

The other major regulatory arena in estuaries centers on wetlands and the policy of mitigation (section 404 provisions of the Clean Water Act of 1972). When wetlands or marshes are removed, filled, or dredged, wetlands must be created, restored, or enhanced in another area within the estuary as a measure of mitigation.

In Oregon, large driftwood is not currently an issue in the mitigation process. The Division of State Lands and the Department of Environmental Quality administer permits that allow the alteration of wetlands and marshes. Although they do not consider driftwood to be an issue in Oregon's tidal marshes, they do not encourage its removal. No credit is given in the mitigation process for either removing or retaining driftwood. In effect, the agencies ignore driftwood because they do not understand the millennial, ecological link of the estuaries to the forest. If healthy estuaries and seas are to exist, however, then driftwood must become part of the mitigation process and, once again, become the creative link between the forest and the sea.

CHAPTER 8

THE RIVERS AND STREAMS

HISTORICAL PERSPECTIVE

Most early descriptions of the streams and rivers of the Pacific Northwest were recorded in journals of British citizens and personnel of the U.S. Army. These descriptions tell of valleys so wet that early travel was confined to hills along their edges. Peter Skene Ogden, of the British-owned Hudson's Bay Company, noted in the early 1800s that much flooding resulted from the beaver dams, sediment accumulation, fallen trees, and live vegetation within stream channels.[23]

Streams in the lowlands around Puget Sound, Washington, were similar to those of the Willamette Valley, Oregon, consisting of a network of sloughs, islands, beaver ponds, and driftwood dams but no discrete channel. The Skagit River lowlands, Washington, once encompassed almost 200 square miles, of which over 50 square miles were marshes, sloughs, and wet meadows. Maps in Reports of the Secretary of War for the years 1875 to 1899 show the lower Nooksack and Snohomish rivers in Washington to have been large areas of sloughs, swamps, and grassy marshes before 1900.[19] All the coastal valleys in Oregon contained many marshes and sloughs. Streams and floodplains in the lowlands of both states had far greater ecological interactions before they were cleared and channeled to accommodate agricultural activity.

Channels of both high- and low-gradient rivers contained large amounts of driftwood, regardless of the nature of the channel bottoms. In

1826, Peter Skene Ogden saw the lower Siuslaw River and the North Fork Siuslaw River, in Oregon, so filled with fallen trees and driftwood that trappers could explore but little of them. In 1854, the Willamette River flowed in five separate channels between the towns of Eugene and Corvallis (Figure 8.1).

Reports of the Secretary of War for the years 1875 to 1921 discussed the many obstacles to navigation in the Willamette River above Corvallis, where the riverbanks were heavily timbered for half a mile on either side (see Figure 3.2). Over 5500 drifted trees were pulled from a 55-mile stretch of the river in a ten-year period (Figure 8.2). These trees ranged from five to nine feet in diameter and from 90 to 120 feet in length. The river was later confined to one channel by the activities of engineers.[19]

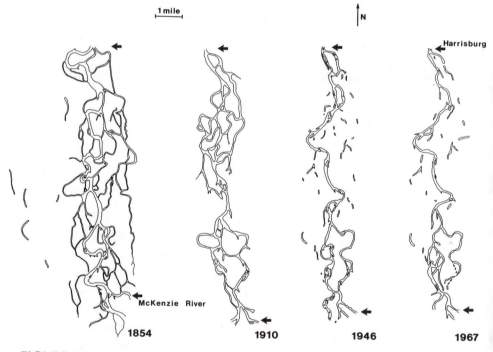

FIGURE 8.1

The Willamette River flowed in several separate channels in 1854. Channels were both formed and abandoned by the river because of the large size and number of trees that clogged them. Boat navigation and development of agriculture on the floodplain resulted in the disappearance of the floodplain forest and the conversion of the Willamette into a single channel. (Reprinted with permission from Sedell, J.R. and J.L. Froggatt. 1984. *Verh. Internatl. Verein. Limnol.* 22:1828–1834.)

FIGURE 8.2
The snagboat *Mathloma* cleared the Willamette River in the late 1890s and helped to keep it confined in a single channel. (Photograph courtesy of the U.S. Army Corps of Engineers.)

As late as 1914, I.A. Williams, a soils scientist at Oregon State Agricultural College in Corvallis, described the condition of Willamette Valley streams as they had been in 1910:

> Many of the smaller streams...through these flat sections of the valley flow sluggishly and frequently overflow their banks during periods of heavy winter rainfall...Most of these have sufficient grade to carry even more water than ordinarily comes to them; seldom less than 3, and usually more, feet of fall per mile. The annual overflow is caused from the obstructing of the channel by the growth of trees and the extension of their roots, the dams thrown across the channels by beavers and the consequent accumulation of sediment and other debris, etc....It is common condition, however, and usually all that is necessary is a clearing out and opening up of the

clogged channel of the stream to afford entire relief...to the farmer....[24]

In both Oregon and Washington, other rivers were completely blocked by driftwood in their lower main channels. The Skagit River, Washington, for example, contained a driftwood jam three fourths of a mile long and one fourth of a mile wide. The Stillaguamish River, Washington, was closed by six driftwood jams from the head of tide-water to river mile seventeen. Drifted trees were so numerous, so large, and so deeply imbedded in the river's bottom that a steam-driven snag boat operated for six months to open a channel only a hundred feet wide.

Captain Hall, a riverboat captain on the Mississippi in the early to mid-1800s, as quoted by geologist Sir Charles Lyell, gave a good description of the perils of navigation caused by such imbedded driftwood:

> Unfortunately for the navigation of the Mississippi, some of the largest trunks, after being cast down from the position on which they grew, get their roots entangled with the bottom of the river, where they remain anchored, as it were, in the mud. The force of the current naturally gives their tops a tendency downwards, and by its flowing past, soon strips them of their leaves and branches. These fixtures, called snags or planters, are extremely dangerous to the stream-vessels proceeding up the stream, in which they lie like a lance in rest, concealed beneath the water, with their sharp ends pointed directly against the bow of vessels coming up. For the most part these formidable snags remain so still, that they can be detected only by a slight ripple above them, not perceptible to inexperienced eyes. Sometimes, however, they vibrate up and down, alternately showing their heads above the surface and bathing them beneath it.[14]

Of such imminent danger were these snags, wrote Lyell, that almost all boats on the Mississippi were constructed on a particular plan to guard them against fatal accidents. He goes on to say that between 1855 and 1865 "the greater number of these trunks of trees have been drawn out of the mud" with the aid of a steam-powered snag boat.

In addition to driftwood stuck in the muddy channels of valley-bottom rivers, driftwood jams in high-gradient rivers were often located where the channel gradient decreased abruptly. A section of the South Fork Nooksack River was thus described in 1883:

> ...we came to a place where the river, during freshets had ground sluiced all the earth away from the roots of the trees, and down some 6 feet to the gravel. This covered a region of country a mile in width by five in length. Overgrown yellow fir timber had once covered most of that section. If the river below there was only clear of jams that place would be paradise for hand loggers. On the gravel lay many million feet of sound fir timber, which only needed to be junked up during the summer and the winter freshets would float the logs down to the sea. Immediately below this place, the jams first extend clear across the river, and for the next 20 miles there is a jam across the river nearly every mile.[20]

Streams and rivers historically supplied estuaries and beaches with wood, the majority of which was transported downstream during high flows. In 1890, a captain in the U.S. Army Corps of Engineers made the following comment, apparently in frustration: "The snagging and bar scraping done on the [Coquille] was productive of great good to the navigation of the river. The freshets of last winter and spring...have, however, again deposited many snags on the river which should be removed."[19]

Streams historically replenished annual supplies of driftwood to the lower portions of river basins and out into the sea, where it washed up on beaches. But the banks of lower rivers and estuaries—the riparian corridor—were probably the common source of large driftwood in the bays.

In the lower Tillamook River, Oregon, in 1904, for example, the U.S. Army Corps of Engineers reported cutting down "all overhanging trees" along the estuary's banks in an attempt to keep a step ahead of the driftwood's accumulation.[25] These corridors were lined with overhanging hardwood trees, which historian Orville Dodge described beautifully in 1898: "For a portion of its course below Myrtle Point the Coquille is fringed with [maple and California laurel] and, when the white man arrived on the scene, in places their tops met and interlaced above the stream. Travel upon the Coquille is through scenes of enchantment, and the sluggish river seems like dim isles in ancient cathedrals."[26]

The woody vegetation along this section of the Coquille and along

FIGURE 8.3
Lowland riverside forests once supplied much driftwood to the ocean. Agricultural activities have reduced such forests to small, fragmented patches and a thin line. (Photograph by Jean Paul Bravard.)

lower sections of other rivers was extensively logged in the 1800s. Upstream, the riverside forests were among the first to be cut, because logs could be floated downriver to ports when no other transportation was available. Major sources of driftwood for estuaries and beaches along the northwest coast were thus destroyed by about 1920 (Figure 8.3).

Substantial amounts of driftwood must have been transported to the sea at the time when most riparian zones were dominated by such large coniferous trees as Douglas-fir, western redcedar, and Sitka spruce and such deciduous trees as black cottonwood, bigleaf maple, Oregon ash, and red alder. Hundreds of millions of board feet of logs and driftwood have entered Puget Sound and Georgia Strait from the rivers draining the Cascade Mountains of Washington and the coastal mountains of British Columbia. They were joined by large numbers of "escapees" from log rafts.

Over 10 billion board feet of logs are annually stored or travel in the estuaries and the lower segments of rivers in the Pacific Northwest and British Columbia, Canada. A one percent escape rate would allow over a hundred million board feet of driftwood to enter the ocean from this source alone.

One can conservatively estimate that in days past as much as two to four billion board feet of wood per year was transported to the sea. Two billion board feet per year is a small amount when prorated across the entire North Pacific. Large driftwood, an important ecological component of Pacific Northwest streams and rivers, interfered with human objectives (Figure 8.4), however, and was summarily removed. In fact,

FIGURE 8.4
River of wood from the forest to the mills. Most rivers had to be cleaned and improved to carry such volumes. (Photograph courtesy of the Forest History Society, Durham, N.C.)

FIGURE 8.5

Logs in a small stream awaiting enough water to be floated to the mill. (USDA Forest Service photograph from historical files.)

people throughout North America have systematically cleaned driftwood from streams and rivers for over 150 years.

From the 1800s to around 1915, streams and small rivers were cleaned of driftwood so that logs could be floated from the forests to the mills (Figure 8.5). Many streams had several splash dams built on them to temporarily augment the flow of water in order to float logs to mills (Figure 8.6). The net effect of channel clearance and splash damming was to remove large quantities of driftwood from medium to large streams, a significant change from the conditions that formerly existed.

These changes were not unique to the Pacific Northwest:

> The rich lumber industry that exploited the primitive coniferous forests of the upper Mississippi basin and the Great Lakes area during the last two thirds of the nineteenth century depended for its existence upon suitable water transportation....On the Chippewa and upper Mississippi rivers

FIGURE 8.6
Splash dam at Austin Place, Hamilton Creek, Oregon, August 16, 1907. Such dams
were used to regulate water flow and float logs to sawmills. (Historical photograph,
courtesy of the Horner Museum, Oregon State University.)

[sic] the contest attained unusual importance because it involved the Mississippi River Logging Company, probably in
its time the largest single aggregation of logging capital in the
world.

The Chippewa River rises in northwestern Wisconsin some
thirty miles south of Lake Superior and flows in a general
southwesterly direction until it empties into the Mississippi at
the foot of Lake Pepin almost directly opposite Wabasha, Minnesota. While certainly not the most navigable, it seems to have
been the richest logging stream of all the tributaries of the
Mississippi. Its wealthy valley, estimated in 1840 to hold one
sixth of all the white pine west of the Alleghenies, attracted the
early and persistent attention of the lumber industry.[27]

In fact, the sentiments of the time were: "We'll never get the pine cut if we log till hell freezes."[28] Streams that were shallow and could not float logs, except during the spring floods, were already being altered so as to accommodate as many logs as possible as early as the mid-1800s in such places as Michigan, Wisconsin, and Minnesota. Others were even less readily suitable for driving logs:

> Levis Creek, a small Wisconsin stream, was choked with brush and lined with alders on the hairpin turns. Before logs could be driven on it, the channel had to be cleared and canals cut across the bends. Even then, logs could not be floated on the ordinary stage of water, but only during rises, and men had to be stationed on both banks of the stream to keep the logs moving....
>
> Dams increased in numbers, becoming more permanent and more expensive to build. By the late eighties [1880s] there were 60 to 70 dams on the Saint Croix River and its tributaries, 41 dams in the Menominee Valley, and some 25 dams on the 95 mile-long Red Cedar River. Dams costing from $6,000 to $16,000 had been built, and $100,000 dams were being planned. Added burdens were incurred a few years later, when Wisconsin taxed logging dams 2 percent of their gross earnings.[29]

Before the great ecological value of driftwood was known, west coast fishery managers believed that driftwood in streams restricted fish passage, supplied material for driftwood jams, and caused channels to scour during floods (Figure 8.7). Indeed, during times of flooding such fears might have seemed to be well-founded, but it is now known that results of stream cleaning have been ecologically disastrous.

CURRENT OUTLOOK

Today, driftwood removal from Pacific Northwest streams and rivers is not practiced on the same scale as in the 1950s and 1970s. Most salmon enhancement programs, over much of the western United States and Canada, are replacing wood by direct reintroduction or by leaving large trees alongside streams for future input. The combination of removing driftwood for fish passage in headwaters of drainage basins, early-day splash damming to transport logs from forests to mills, and removing driftwood from large rivers for navigation has created a situation in

FIGURE 8.7
Stream channel scoured by debris torrent that no longer carries large driftwood into larger streams. (USDA Forest Service photograph by Fred. J. Swanson.)

which entire drainage systems now contain only a feeble fraction of the physical amount and ecological function of the driftwood—such as salmon habitat—once present in pristine streams and rivers.

Logging causes episodic inputs of woody debris into streams. Logging slash incidentally introduced into a stream channel during logging and large trees, which accidentally slide into the stream from a steep hillside, may become debris in the channel (Figure 8.8).

FIGURE 8.8
Clearcut on steep slopes often puts small logging slash into streams. (Photograph courtesy of the USDA Forest Service.)

Following logging, concentrations of woody debris in small, steep, western Oregon streams may increase fourfold above levels existing before logging. The amount of debris that enters such a stream depends on the method of logging and whether or not a buffer strip is present to protect the stream (Figure 8.9). Additional woody debris may enter streams when high water moves logging slash into a channel after the channel has been cleaned.

Log rafts, pilings, and wooden bridges are major sources of wood in many of today's large streams and rivers (Figure 8.10). Tens of millions of board feet of logs have been rafted on almost all of the large Pacific Northwest rivers. Two to five percent of the rafted logs sank before arriving at their final destination. Even with this small loss, the vast numbers of sunken logs have added a great supply of wood to lower portions of coastal rivers.

Millions of pilings have been driven into the bottoms of large rivers and coastal streams above tide water (Figure 8.11). These structures trap small branches and pieces of driftwood that provide important fish habi-

FIGURE 8.9
At present, most wood getting into streams is small slash created during logging. (Photograph by Peter Bisson.)

FIGURE 8.10
Log rafts and pilings in large rivers and estuaries provide habitat for birds and fish and a supply of driftwood. (Weyerhaeuser Co. photograph, courtesy of the Forest History Society, Durham, N.C.)

FIGURE 8.11
A buffer strip that will provide driftwood to the stream in the future. (USDI Bureau of Land Management photograph by John Anderson.)

tat. Further, numerous, large coastal streams in the Pacific Northwest were once bridged by filling the channel with large logs. Many of these solid log bridges or crossings still exist, as do the wooden abutments for more conventional bridges built 20 to 40 years ago.

In comparing today's relatively pristine streams with those in logged areas, it is seen that the average length and frequency of large driftwood is far greater in the most pristine streams. For example, in the Oregon Coast Range, one small stream drained an unlogged basin in a wilderness area, and another drained a basin outside the wilderness area that had been about 95 percent logged during the last 30 years. The total number of pieces of wood and their average diameters and lengths were determined.

The stream draining the wilderness area had more than ten times the amount of driftwood per given distance in the stream's channel than did the stream draining the logged area. Whereas the average diameter of wood in both streams was about 20 inches, the average length of wood in the stream draining the wilderness was about 33 feet as opposed to 16 feet in the logged area.[30]

Another case study was conducted along the North Fork Breitenbush, which is a high-gradient, sixth-order stream in the central Cascade Range of Oregon. The stream's width when full to the highest edges of its banks is often greater than 98 feet, large enough to move even the largest and the longest drifting tree. The stream, divided by a wilderness area boundary, had been logged outside the wilderness area several times since 1965 to remove the dying and dead trees. Logging had taken place on the lower 60 percent of its fish-bearing channels.

Approximately 67 percent of the driftwood in the wilderness with attached rootwad was longer than 49 feet, as opposed to 23 percent in the logged area. Stumps made up over 60 percent of the wood in the logged area, as opposed to less than 6 percent in the wilderness (Figure 8.12).

FIGURE 8.12
Stump frequency along a stream is a good indicator of human-caused uncoupling of forest and stream, as well as an indicator of driftwood-related habitat lost to the stream. (USDA Forest Service photograph by James R. Sedell.)

Although the total volume of wood was the same, the wood's stability was drastically reduced in the logged portion, resulting in a 40 percent widening of the stream's channel.[31] Clearly, the trend is toward smaller, less frequent, and less stable driftwood in logged streams, which results in wider, less stable stream channels.

The state of Washington, for example, has required thorough removal of logging slash from streams from the 1970s into the 1980s. During the same time, the effect of stream cleaning has been evaluated. In almost every case, removing driftwood has resulted in the loss of important habitat features and thus a loss in species richness of fishes. Removing large, stable wood along with smaller material reduces the stability of channels and the quality of cover and habitat in pools.

Small, easily floated driftwood has occasionally become temporarily abundant after a stream has been cleaned. Such increases in small driftwood have come from new slash entering the water, either because it was stacked near the high-water line or because driftwood already in the stream was broken up prior to logging. In most cases, unstable driftwood is flushed from a cleaned reach of stream within a few years, and the channels are diminished of structural diversity.

In cleaned stream channels, the proportion of riffles increases and the number of pools decreases. The increased frequency of riffles increases the abundance of steelhead and cutthroat trout under one year old, which prefer riffles as habitat, but the abundance of coho salmon, steelhead, and cutthroat trout of older age classes, which prefer pools, is correspondingly reduced.

State and federal agencies in the 1970s and 1980s have annually spent over six million dollars to remove driftwood jams, but the impact of this action on habitat and how it functions had been little examined until recently. Populations of fish in the vicinity of driftwood jams were reduced immediately after the jams were removed, but no significant decline in abundance was detectable where part of a jam was allowed to remain.[32]

In addition, areas of spawning and rearing downstream from the old driftwood jam were damaged by large sediment flows that had been stabilized upstream from the removed jams. Partial removal of a driftwood jam appears to be biologically and economically preferable to complete removal.[33]

Salvage logging in streams began in the Pacific Northwest after the great storm of 1964, a magnitude of storm that occurs about once a century. Since then, the federal government has funded cleaning and wood salvage from streams after every major storm. The focus of such activities tended to be timber sales on both public and private lands adjacent to streams of fifth order or larger because of available access to established main roads and drainage improvement on agricultural lands. The main reason cited for salvaging driftwood in streams is to protect bridges and culverts downstream of the driftwood in order to reduce cases of property liability in court.

Over the last hundred years, millions of drifted trees and other driftwood have been cleared out of streams and rivers throughout North America to facilitate navigation and reduce flooding. To this end, streams and rivers have been channelized and dammed and marshes have been drained. In addition, most stream banks have been so altered through agriculture, livestock grazing, roading, logging, and mining that they now have dramatically smaller and younger trees of different species than in times past (Figure 8.13).

FIGURE 8.13
Clearcutting into stream bottoms has completely altered riparian areas and the kinds and sizes of trees available for driftwood. (Photograph courtesy of the USDA Forest Service.)

Most big western redcedars and Douglas-firs have been logged along Cascade Mountain streams and along coastal streams of greater than third order. On private land, more than 70 percent of the coniferous trees greater than 14 inches in diameter at breast height have been logged within a hundred feet of fish-bearing streams.

Even today, county sheriffs, port commissions, and recreational boaters still routinely clear driftwood from rivers for safety and convenience. As a result, most Pacific Northwest streams and rivers, as well as the rest of the world's streams and rivers in economically developed countries, bear little resemblance to their ancestral versions, when they flowed freely through pristine forests carrying their gift of driftwood to the sea.

In light of recent ecological information, it is imperative to re-examine the way streams and rivers are treated and the supply of driftwood they no longer carry on its journey to the sea. The present course, which is helping to make extinct the runs of wild salmon through the continuing loss of large driftwood in their habitat, leads toward impoverishment of the waterways and ultimately the ocean. There are, however, alternatives. This self-destructive path can be redirected. The choice is ours. To the children we bequeath the consequences. How shall we choose?

APPENDIX

COMMON AND SCIENTIFIC NAMES OF PLANTS AND ANIMALS

Common name	Scientific name
PLANTS	
Trees and shrubs	
Beech	*Fagus* spp.
Bigleaf maple	*Acer macrophyllum*
Birch	*Betula* spp.
Black cottonwood	*Populus trichocarpa*
California laurel	*Umbellularia californica*
Coast redwood	*Sequoia sempervirens*
Douglas-fir	*Pseudotsuga menziesii*
Fir	*Abies* spp.
Hemlock	*Tsuga* spp.
Maple	*Acer* spp.
Oak	*Quercus* spp.
Oregon ash	*Fraxinus latifolia*
Pine	*Pinus* spp.

Common name	Scientific name

PLANTS
Trees and shrubs (continued)

Red alder	*Alnus rubra*
Silver fir	*Abies amabilis*
Sitka alder	*Alnus sinuata*
Sitka spruce	*Picea sitchensis*
Spruce	*Picea* spp.
Streamside sage	*Artemisia ludoviciana*
Subalpine fir	*Abies lasiocarpa*
Sugar pine	*Pinus lambertiana*
Willow	*Salix* spp.

Mosses

Moss	Bryophyta

Horsetails

Horsetail	*Equisetum* spp.

Grasses and grass-like plants

Eelgrass	*Zostera* spp.
European beachgrass	*Ammophila arenaria*
Grass	Graminae

Herbs

Alpine shooting star	*Dodecatheon alpinum*
Beargrass	*Xerophyllum tenax*
Clover	*Trifolium* spp.
Fireweed	*Epilobium* spp.
Glacier lily	*Erythronium grandiflorum*
Stone crop	*Sedum* spp.
Subalpine lupine	*Lupinus latifolius*
Thistle	*Cirsium* spp.
White marsh marigold	*Caltha biflora*

Common name	Scientific name
INVERTEBRATES	
Clams and mussels	
Chiton	Amphineura
Deep-sea wood-borers	*Xylophaga* spp.
Introduced shipworm	*Teredo navalis*
Native shipworm	*Bankia setacea*
Washington deep-sea wood-borer	*Xylophaga washingtona*
Crustaceans	
Barnacle	Cirripedia
Copepod	Copepoda
Northern gribble (native)	*Limnoria lignorum*
Southern gribble	*Limnoria tripunctata*
Insects	
Alderfly	Sialidae
Ancient dragonfly	*Tanypteryx hageni*
Beetle	Coleoptera
Coast tiger beetle	*Cicindela bellissima*
Cranefly	Tipulidae
Fly	Diptera
Hairy caddisfly	*Cryptochia pilosa*
Mayfly	Ephemeroptera
Midge	Chironomidae
Mosquito	Culicidae
Oceanic water strider	*Halobates* spp.
Pond strider	*Gerris* spp.
Riffle beetle	*Lara avara*
Stonefly	Plecoptera
True bug	Hemiptera
Wood-eating cranefly	*Lipsothrix fenderi, L. nigrilinea*

Common name	Scientific name
INVERTEBRATES	
Snails	
Slug	Pulmonata
Snail	Pulmonata
Spiders and allies	
Spider	Araneida
Worms	
Polychete worms	Polychaeta
Roundworm	Nematoda
Jellyfish	
By-the-wind sailors	*Velella* spp.
Portuguese man-of-war	*Physalia* spp.
VERTEBRATES	
Fish	
Bigeye tuna	*Thunnus obesus*
Chinook salmon	*Oncorhynchus tschawytscha*
Chum salmon	*Oncorhynchus keta*
Coho salmon	*Oncorhynchus kisutch*
Cutthroat trout	*Salmo clarki*
Dolphin fish (mahimahi)	*Coryphaena hippurus*
Dorado	*Coryphaena* spp.
Large-scale sucker	*Catostomus macrocheilus*
Mountain whitefish	*Prosopium williamsoni*
Pacific herring	*Clupea harengus*
Sculpin	*Cottus* spp.
Silky shark	*Carcharhinus falciformis*
Skipjack tuna	*Katuwouvs pelamis*
Starry flounder	*Platichthys stellatus*
Steelhead trout	*Salmo gairdneri*

Common name	Scientific name

VERTEBRATES
Fish (continued)

Stickelback	*Gasterosteus aculeatus*
Sturgeon	*Acipenser* spp.
Trout	*Salmo* spp.
Whale shark	*Rhincodon typus*
Yellowfin tuna	*Thunnus albacares*

Amphibians

Pacific giant salamander	*Dicamptodon tenebrosus*
Tailed frog	*Ascaphus truei*

Reptiles

Sea turtles	Chelonidae, Dermochelidae
Snakes	Ophidia

Birds

American crow	*Corvus brachyrhynchos*
American robin	*Turdus migratorius*
Bald eagle	*Haliaeetus leucocephalus*
Brown pelican	*Pelecanus occidentalis*
Cedar waxwing	*Bombycilla cedrorum*
Common merganser	*Mergus merganser*
Common nighthawk	*Chordeiles minor*
Common raven	*Corvus corax*
Cormorant	*Phalacrocorax* spp.
Golden eagle	*Aquila chryaetos*
Great blue heron	*Ardea herodias*
Great egret	*Casmerodius albus*
Green heron	*Butorides striatus*
Gull	*Larus* spp.
Harlequin duck	*Histrionicus histrionicus*
Song sparrow	*Melospiza melodia*

Common name	Scientific name

VERTEBRATES
Birds (continued)

Spotted sandpiper	*Actitis macularia*
Steller jay	*Cyanocitta stelleri*
Swainson thrush	*Catharus ustulatus*
Varied thrush	*Ixoreus naevius*
Violet-green swallow	*Tachycineta thalassina*
Water ouzel	*Cinclus mexicanus*
Winter wren	*Troglodytes troglodytes*

Mammals

Beaver	*Castor canadensis*
Black bear	*Ursus americanus*
California bat	*Myotis californicus*
Deer mouse	*Peromyscus maniculatus*
Dolphins	Delphinidae
Dusky shrew	*Sorex monticola*
Harbor seal	*Phoca vitulina*
Hare	*Lepus* spp.
Long-tailed vole	*Microtus longicaudus*
Long-tailed weasel	*Mustela frenata*
Marsh shrew	*Sorex bendirei*
Marten	*Martes americana*
Mink	*Mustela vison*
Northern flying squirrel	*Glaucomys sabrinus*
Oregon red tree vole	*Arborimus longicaudus*
Rabbit	Lagomorpha
Red-backed vole	*Clethrionomys* spp.
Red fox	*Vulpes vulpes*
Seals	Otariidae/Phocidae
Sea lions	Otariidae
Short-tail weasel	*Mustela erminea*

Common name	Scientific name

VERTEBRATES
Mammals (continued)

Common name	Scientific name
Spotted skunk	*Spilogale putorius*
Squirrel	Sciuridae
Townsend chipmunk	*Eutamias townsendi*
Weasel	*Mustela* spp.
Western red-backed vole	*Clethrionomys californicus*

GLOSSARY

Abrasion: the process of wearing down or rubbing away by means of friction

Abyssal: of or pertaining to the great depths of the oceans

Aerobic: said of an organism, such as a bacterium, that can only live or function in the presence of free oxygen; also said of processes or conditions that can only occur in the presence of free oxygen

Aggregation: a collection of material at a given place

Alga (pl. algae): any of various primitive, chiefly aquatic, one-celled or multicellular plants that lack true stems, roots, and leaves but usually contain the green pigment called chlorophyll

Alluvial material: refers to the material that is deposited by flowing water

Anaerobic: said of an organism, such as a bacterium, that does not require free oxygen to live or function; also said of processes or conditions that do not require oxygen

Ancient forest: a forest that is past full maturity; the last stage in forest succession; a forest with two or more levels of canopy, heart rot, and other signs of obvious physiological deterioration

Angiospermae: any of a class of plants that is identified by having their seeds enclosed in an ovary

Antenna (pl. antennae): a pair of segmented appendages located on the head above the mouthparts, which are usually sensory in function

Aquatic: an organism that lives in water

Backwater: water held or pushed back by or as if by a dam or current; especially a body of stagnant or still water thus formed

157

Bacterium (pl. bacteria): any of numerous one-celled microorganisms of the class Schizomycetes, occurring in a wide variety of forms, existing as free-living organisms or as parasites

Bar: anything that impedes or prevents; an obstacle, such as a sandbar in a river that impedes the flow of water

Bark: the outer, protective layer of woody branches, stems, and roots of trees and other woody plants

Bay: a body of water partly enclosed by land that has a wide outlet to the sea

Beach: the shore of a body of water, especially when sandy or pebbly; the sand or pebbles on a shore between the line of high tide and the first, permanent terrestrial plant community; (*v.*) to strand on a beach, such as driftwood that is blown ashore during a storm and is stranded on the beach

Beached: referring to driftwood or some other object that is washed onto a beach during high water and that is stranded on the beach when the water recedes

Bedload: soil, rocks, and other debris rolled along the bottom of a stream by moving water, in contrast to siltload, which is carried in suspension

Bedrock: the solid rock that underlies soil, sand, clay, gravel, and loose material on the earth's surface

Biomass: the combined weight of all living organisms in a given area

Biotic: composed of plants and animals

Bivalve: a mollusc, such as an oyster or a clam, having a shell that consists of two hinged halves

Board foot: a unit of measure of lumber that is equal to one foot square by one inch thick

Brackish: containing some salt; briny

Buffer strip: a strip of vegetation that is left along the margins of a stream or a river to reduce the impact of logging activities on that stream or river

Calcareous: composed of, containing, or characteristic of calcium carbonate, calcium, or limestone; chalky

Calcium: a silvery, moderately hard, metallic element that constitutes approximately three percent of the earth's crust and is a basic component of bone, shell, and leaves

Calcium carbonate: a colorless or white crystalline compound that occurs naturally in chalk, limestone, marble, and other forms

Carbon: a naturally abundant non-metallic element that occurs in many inorganic and in all organic compounds

Carnivore: a mammal that eats the flesh of other animals

Cascade: a waterfall or a series of small waterfalls over steep rocks

Cecum: the large, blind pouch that forms the beginning of the large intestine

Cellulase: an enzyme that digests cellulose; *see* Cellulose

Cellulose: an amorphous carbohydrate polymer, the main constituent of all plant tissues and fibers, that is used in the manufacture of many fibrous products, including paper, textiles, and explosives

Cellulosic: composed of cellulose

Channel: the bed of a stream or river

Chitin: a semitransparent, horny substance that forms the principal component of the shells of crustaceans, the exoskeletons of insects, and the cell walls of certain fungi

Clay: a very fine-grained sediment that becomes plastic and acts like a lubricant when wet; consists primarily of hydrated silicates of aluminum and is widely used in making bricks, tiles, and pottery

Climatic cycle: the cyclic changes in weather patterns in a geographical area over time

Colon: the section of the large intestine that extends from the cecum to the rectum

Colonization: the process or act of establishing a colony or colonies; *see also* Colony

Colony: a group of the same kind of plants or animals living together

Community: a group of one or more populations of plants and/or animals using a common area; an ecological term used in a broad sense to include groups of plants and animals of various sizes and degrees of integration

Conifer: the most important order of the Gymnospermae, comprising a wide range of trees, mostly evergreens that bear cones and have needle-shaped or scale-like leaves; its timber is commercially identified as softwood

Coniferous: of or pertaining to conifers

Coniferous forest: a forest dominated by cone-bearing trees; *see also* Conifer

Continental shelf: a generally shallow, flat, submerged portion of a continent that extends to a point of steep descent to the floor of the ocean; generally the most productive part of the sea

Continuum: a continuous extent, succession, or whole, no part of which can be distinguished from neighboring parts except by arbitrary division

Copepod: any of numerous small marine and freshwater crustaceans of the order Copepoda

Creep: very slow downslope movement of soil resulting from deformation of soil under the influence of gravity; rate of speed is fractions of an inch per year

Crown: the upper part of a tree or other woody plant that carries the main branching system and foliage and that surmounts at the crown's base a more or less clean stem

Crustacean: any of various predominantly aquatic arthropods of the class Crustacea, including lobsters, crabs, shrimps, and barnacles, characteristically having a segmented body, a chitinous exoskeleton, and paired, jointed limbs

Debris flow: very fast movement of water-charged soil and vegetation, both dead and alive, down stream channels; rate of speed is in feet per second

Debris slide: very fast movement of soil, commonly containing a high concentration of water, down hillslopes under the influence of gravity; the initial sliding surface is generally 3 to 12 feet below the surface of the ground

Debris torrent: a slurry of debris, soil, and water that occurs where a landslide enters a steep channel and moves downstream during conditions of severe flooding

Decay: to decompose; to rot; in wood, the decomposition by fungi or other microorganisms results in softening, progressive loss of strength and weight, and changes in texture and color

Deciduous: pertaining to any plant organ, such as a leaf, that is shed naturally; also referring to perennial plants that shed their leaves and are therefore leafless for some time during the year

Delta: the usually triangular deposit of alluvial material at the mouth of a river

Detritivore: an organism that feeds on decaying organic materials

Detritus: any disintegrating organic material or debris

Diatom: any of various minute, one-celled or colonial algae of the class Bacillariophyceae, having siliceous cell walls consisting of two overlapping, symmetrical parts

Diversity: the relative degree of abundance of species of plants and animals, functions, communities, habitats, or habitat features per unit of area

Dynamic: characterized by or tending to produce continuous change

Earthflow: slow downslope movement of earth material under the influence of gravity; rate of speed is in inches to feet per year; the moving mass is generally between 10 to 70 feet thick; the sliding surface is much thinner, however, and may be flat or irregular in form

Ecological: referring to a relationship between living organisms and their non-living, physical environment

Ecosystem: all the living organisms interacting with their non-living, physical environment, considered as a unit

Eddy (pl. eddies): a current, as of water or air, moving contrary to the direction of the main current, especially in a circular motion

El Niño: a regional or global oceanic–atmospheric disturbance whose manifestations range from increased sea surface temperature in the tropical east Pacific to aberrant rainfall patterns

Embayment: a bay or baylike shape; the formation of a bay; *see* Bay

Enzyme: any of numerous proteins produced by living organisms and functioning as biochemical catalysts that break down materials, such as ingested food, in living organisms

Epipelagic: an ocean's surface layer

Erosion: the group of processes, including weathering, dissolution, abrasion, corrosion, and transportation, by which earthy or rock material is removed from any part of the earth's surface

Estuarine: of an estuary

Estuary: the wide, lower course of a river where its current is met and influenced by the tides; it also is defined as an arm of the sea that extends inland to meet the mouth of a river

Fecal material: material discharged from the bowels; more generally, any discharge from the digestive tract of an organism

Filament: a slender threadlike structure

Filamentous: like a filament

First-order stream: *see* Stream order

Floodplain: a plain bordering a stream or river that is subject to flooding

Flora: plants collectively, especially the plants of a particular region or time

Flotsam: material floating on the surface of the water that tends to collect in eddies above debris dams

Foliage: the leaves of growing plants; plant leaves collectively

Forage: vegetation used for food by wildlife, particularly wild ungulates, such as deer and elk

Forb: any herbaceous species of plant other than grasses, sedges, or rushes; fleshy-leaved plants

Foredune: a vegetated sand dune immediately inland from the beach that parallels it

Forest: generally, that portion of the ecosystem characterized by tree cover; more particularly, a plant community predominantly of trees and other woody vegetation that grow close together

Forest floor: the surface layer of a soil that supports forest vegetation

Fourth-order stream: *see* Stream order

Freshet: the sudden overflow of a stream or river resulting from a heavy rain, or thaw, or both

Function: the natural or proper action for which an organism, habitat, or behavior has evolved

Fungal: caused by or associated with fungi; *see also* Fungi

Fungal hypha (pl. hyphae): *see* Fungus; Hypha

Fungus (pl. fungi): mushrooms, truffles, molds, yeasts, rusts, etc.; simply organized plants, unicellular or made of cellular filaments or strands called hyphae, lacking chlorophyll; fungi reproduce asexually and sexually with the formation of spores

Fur: the thick coat of soft hair covering the body of any of various mammals, such as beaver, fox, or cat

Genus: the first word in a binomial or scientific name

Geomorphology: the geological study of the configuration and evolution of land forms

Germinate: to begin to grow, to sprout

Gland: an organ that extracts specific substances from the blood and concentrates or alters them for subsequent secretion

Glycoside: any of a group of organic compounds, occurring abundantly in plants, that produce sugars and related substances on hydrolysis

Gradient: a rate of inclination, a slope; an ascending or descending part; an incline

Grass: any species of plant that is a member of the family Graminae, characteristically having narrow leaves, hollow, jointed stems, and spikes or clusters of membranous flowers borne in smaller spikelets; such plants collectively

Gymnospermae: a group of woody plants having a naked seed, i.e., a seed not enclosed in an ovary

Habitat: the sum total of environmental conditions of a specific place occupied by a plant or animal, or a population of such species

Habitat niche: the peculiar arrangement of food, cover, and water that meets the requirements of a particular species

Hardwood: the wood of broad-leaved trees, and the trees themselves, belonging to the botanical group Angiospermae; distinguished from softwoods by the presence of vessels

Headland: a point of land, usually high and with a sheer drop, extending out into a body of water; a promontory

Hemoglobin: a red, respiratory pigment in the blood that contains iron and aids the blood cells in the transport of oxygen to the various parts of the body

Herb: a non-woody plant as distinguished from a woody plant

Herbaceous: pertaining to or characteristic of an herb or non-woody plant as distinguished from a woody plant

Hydraulic: of, involving, moved, or operated by a fluid, especially water, under pressure

Hydraulic complexity: the system of variations in the velocity of the water and the water-operated physical processes in a stream as governed by the physical obstructions within the stream's channel

Hydrolysis: decomposition of a chemical compound by reaction with water

Hydrothermal vent: an opening in the floor of the deep ocean through which issues hot water, often rich in sulfur

Hypha (pl. hyphae): filament or strand of a fungus thallus (non-reproductive vegetative body) that is composed of one or more cylindrical cells; increases by growth at its tip; gives rise to new hyphae by lateral branching

Hypogeous: belowground

Impervious: incapable of being penetrated

Ingest: to take within, to eat

Inorganic: involving neither organic life nor the products of organic life; not composed of organic matter, especially minerals; contrast Organic

Inorganic compound: a chemical compound that does not involve organic products; *see* Inorganic; Organic

Inorganic materials: anything that does not involve organic products; *see* Inorganic; Organic

Integrator: an attribute that expresses the combined influences of a number of interacting variables

Integrity: the state of being unimpaired; soundness; completeness; unity

Intermittent: starting and stopping at intervals, such as a stream whose flow is periodically interrupted

Internal succession: the process of change stimulated primarily by decay and deterioration in a snag or fallen tree

Invertebrate: an animal lacking a backbone or spinal column

Isopod: any of numerous crustaceans of the order Isopoda, which includes the sow bugs and gribbles

Jam: to block, congest, or clog, such as with fallen trees and other woody debris.

Landslide: the dislodging and fall of a mass of earth and rock; *see also* Debris flow; Earthflow

Larva (pl. larvae): the general term for the newly hatched, earliest stage of any of various animals that undergo metamorphosis, such as insects, frogs, and salamanders, and differ markedly in form and appearance from the adult

Lee: the side or quarter away from the direction from which the wind blows; the side sheltered from the wind

Lichen: a plant that is actually two plants in one; the outer plant is a fungus that houses the inner plant, an alga

Lignin: the chief non-carbohydrate constituent of wood; a polymer that functions as a natural binder and support for the cellulose fibers of woody plants

Litter (animal): the number of young produced at a single birth

Litter (forest): the uppermost layer of organic debris on the floor of a forest; essentially the freshly fallen or slightly decomposed vegetable material, mainly foliate or leaf litter, but also twigs, wood, fragments of bark, flowers, and fruits

Litter fall: the fall of litter to the floor of the forest

Logging: the activity of cutting down trees and cutting their stems into predetermined lengths for sale to mills where the stems are converted into lumber or other products

Lookout: a structure used by an animal for a better vantage point

Mammal: an animal that has hair on its body during some stage of its life and whose babies are initially nurtured by their mother's milk

Mandible: the lower jaw of invertebrates

Marine: of the ocean

Meander: to follow a winding and turning course, such as streams that flow through level land; also said of a bend in a stream that has been cut off from the main channel by the stream having shifted its course

Membrane: a thin, pliable layer of tissue that covers surfaces, or separates, or connects regions, structures, or organs of an animal or plant

Membranous: like a membrane; thin, pliable, and more or less transparent; thin and pliable

Metabolic: of, pertaining to, or exhibiting metabolism

Metabolism: the complex of physical and chemical processes involved in the maintenance of life

Metabolite: any of various organic compounds produced by metabolism

Metamorphic: of or relating to metamorphosis; in geology, a rock that has been changed by metamorphism

Metamorphism: any alteration in composition, texture, or structure of rock masses caused by great heat or pressure

Metamorphosis: change in form during development

Microbe: microscopic organism

Microorganism: a plant or animal of microscopic size, especially a bacterium or a protozoan

Microscopic: too small to be seen by the unaided eye, large enough to be seen with the aid of a microscope; exceedingly small; minute

Microtopography: the features of a small place or region, such as one square foot, as opposed to the features of a large place or region, such as one square mile

Mineral: any naturally occurring, homogeneous, inorganic substance that has a definite chemical composition and characteristic crystalline structure, color, and hardness

Mineralize: to convert to a mineral substance

Mineral soil: soil composed mainly of inorganic materials and with a relatively low amount of organic material

Mollusk: any member of the phylum Mollusca, of largely marine invertebrates, including the edible shellfish, such as clams, and some 100,000 other species

Moss: any of various green, usually small plants of the class Musci within the division Bryophyta

Multilayered: a forest canopy with two or more distinct layers of trees

Mycelium (pl. mycelia):   the vegetative part of a fungus, consisting of a mass of branching threadlike filaments or strands called hyphae

Mycorrhiza: the symbiotic relationship of a fungus with the roots of certain plants

Mycorrhizal fungus: a fungus that forms mycorrhiza

Niche: *see* Habitat niche

Nitrogen: a non-metallic element constituting nearly four-fifths of the air by volume, occurring as a colorless, odorless, almost inert gas; it occurs in various minerals and in all proteins

Nitrogen fixation: the conversion of elemental nitrogen (N) from the atmosphere to organic combinations or to forms readily usable in biological processes

Nitrogen-fixing bacteria: bacteria that can take nitrogen gas out of the air and transform it into an organic compound that plants can use

Nocturnal: to be active during the hours of darkness

Null point: that point in an estuary where the velocity of the downstream flow of less dense, less saline water near the surface of the channel and the upstream flow of more dense, more saline water near the bottom of the channel meet and effectively cancel one another out

Obligate: an organism that is able to survive only in a specific type of environment or only in a specific type of relationship within a variety of environments

Omnivore: a animal that eats both plant and animal tissue; a generalized feeder, such as most humans

Omnivorous: of or pertaining to an animal that is an omnivore

Organic: of, pertaining to, or derived from living organisms; of or designating compounds containing the element carbon; contrast Inorganic

Organic combinations: mixtures of organic substances

Organic compound: a chemical compound that involves carbon and is derived from living organisms; *see* Inorganic; Organic

Overflow channel: a channel that is created and used by a stream when it is overflowing its banks during a flood

Oxygen: a colorless, odorless, tasteless gaseous element constituting 21 percent of the atmosphere by volume; it combines with most elements, is essential for plant and animal respiration, and is required for nearly all combustion and combustive processes

Pelagic: of, pertaining to, or living in open oceans or seas rather than waters adjacent to land or inland waters

Photosynthesis: the process by which chlorophyll-containing cells in green plants convert incident light to chemical energy and synthesize organic compounds from inorganic compounds, especially carbohydrates from carbon dioxide and water, with the simultaneous release of oxygen

Phytoplankton: plant organisms, generally microscopic, that float or drift in great numbers in fresh or salt water

Piling: a log driven into the bottom of a stream, river, or estuary to support a structure or to use as a mooring for such things as rafts of logs

Plankton: plant and animal organisms, generally microscopic, that float or drift in great numbers in fresh or salt water

Predaceous: feeding as a predator

Predator: any animal that kills and feeds on other animals

Process: a system of operations in the production of something; a series of actions, changes, or functions that brings about an end or result

Progradation: the process of building upward and outward of land through the deposition of sediments that result from disturbances to the stability of the upslope water catchments, which increases the amount of sediments that are carried down the streams and rivers and thereby increases the amount of sediments that are deposited in a given locale in the lowlands

Protozoan (pl. protozoa): any of the single-celled, usually microscopic organisms of the phylum or subkingdom Protozoa, which includes the most primitive forms of animal life

Protozoon (pl. protozoans or protozoa): any of the single-celled, usually microscopic organisms of the phylum or subkingdom Protozoa, which includes the most primitive forms of animal life

Pupa (pl. pupae): the stage between the larva and the adult in insects with complete metamorphosis; it is a non-feeding and usually outwardly inactive stage

Pupate: to transform to a pupa

Rapid: an extremely fast-moving part of a river, caused by a steep descent in the riverbed

Reach of stream: a stretch of water that is geomorphically defined, such as between bends in a stream, river, or channel or between the start of a meadow or canyon

Recruitment: juvenile individuals of a given species entering a population for the first time

Rectum: that portion of the large intestine extending from the sigmoid flexure to the anal canal

Redd: gravel on the bottom of a stream in which trout and salmon lay their eggs

Resin: any of numerous clear or translucent, yellow or brown, solid or semisolid, viscous substances of plant origin, such as copal, rosin, and amber

Riffle: a stretch of choppy water caused by a shoal or sandbar lying just below the surface of a waterway

Riparian zone: an area identified by the presence of vegetation that requires free or unbound water or conditions more moist than normally found in the area

Riverine: of the river

Rootlet: the small root of a plant

Rootwad: the mass of roots, soil, and rocks that remains intact when a tree, shrub, or stump is uprooted

Salinity: of, relating to, or containing salt; saltiness

Salmonid: a fish of the family Salmonidae, which includes trout and salmon

Salvage logging: the removal and sale of trees that are dead, dying, or deteriorating

Saprophyte: a plant that lives on and derives its nourishment from dead or decaying organic matter

Scale: a small platelike structure of the skin that characteristically forms the external covering of fishes, reptiles, and certain mammals; a similar part, such as one of the minute structures overlapping to form the covering on the wings of butterflies and moths

Scavenger: an animal that feeds on dead plants and/or animals, or on animal wastes

Scientific name: the binomial or two-word latinized name of an organism; the first word describes the genus, the second the species

Sculpin: a small fish

Seamount: a submarine mountain rising to more than 3000 feet above the ocean floor but having a summit at least 1000 feet below the surface

Second-order stream: *see* Stream order

Secrete: to generate from bodily cells or fluids

Secretion: the process of secreting a substance, especially one that is not a product of bodily waste, from blood or cells

Sediment: material suspended in water; the deposition of such material onto a surface, such as a stream bottom, underlying the water

Semiaquatic: adapted for living or growing in or near water; not entirely aquatic

Semiterrestrial: adapted for living or growing on or near land; not entirely terrestrial

Shoal: a place in any body of water where the water is especially shallow

Shrub: a plant with persistent woody stems and relatively low growth form; usually produces several basal shoots as opposed to a single stem; differs from a tree by its low stature and non-treelike form

Side channel: a channel that extends off the main channel of a stream but remains connected to it, such as the backwater of an overflow channel or of a cut-off meander

Silt: a sedimentary material consisting of fine mineral particles intermediate in size between sand and clay

Siltload: silt that is carried in suspension, in contrast to soil, rocks, and other debris rolled along the bottom of a stream by the moving water

Siphon: a tubular organ, especially of aquatic invertebrates, such as squids or clams, by which water is taken in or is expelled

Slash: a collective term used for the roots, branches, tops of the trees, and other unmerchantable fragments that are left on the site after the trees have been cut down and the desirable pieces—logs—have been removed

Slope: the incline of the land surface measured in degrees from the horizontal; also characterized by the compass direction it faces

Slough: a stagnant swamp, marsh, bog, or pond, especially as part of a backwater, inlet, or bayou

Smolt: a young salmon at the state at which it migrates from fresh water to the sea

Snag: a standing dead tree from which the leaves and most of the branches have fallen; such a tree broken off but still more than 20 feet tall; also a large drifted tree stuck in the muddy bottom of a river or estuary that may or may not be visible above the surface

Soil: earth material so modified by physical, chemical, and biological agents that it will support rooted plants

Species: a unit of classification of plants and animals consisting of the largest and most inclusive array of sexually reproducing and cross-fertilizing individuals that share a common gene pool

Splash dam: a small, wooden dam that temporarily retains water that can be released at will to raise the level of the water in a stream's channel to float logs to a larger river

Spring wood: young, usually soft, fast-growing wood with large cells that lies directly beneath the bark and develops in early spring when there is ample moisture; spring wood can be seen in the stump of a tree as the lighter, larger rings in the wood, as opposed to the summer wood, which is represented by the smaller, darker rings in the wood

Stem: the principal axis of a plant from which buds and shoots develop; with woody species, the term applies to principal axes, or trunks, of all ages and thicknesses

Stream order: the classification of the size of a stream based on its volume of water; headwater stream channels are designated first order; two first-order streams combine to form a second-order stream; two second-order streams combine to form a third-order stream, and so forth

Summer wood: wood that develops during the latter part of the growing season when the supply of water is not so ample as in spring and therefore has smaller cells, which in turn makes it darker, harder, and less porous than spring wood; summer wood can be seen in the stump of a tree as the darker, smaller rings in the wood, as opposed to the spring wood, which is represented by the larger, lighter rings in the wood

Suspended sediments: sediments that are suspended in and by the water in which they are carried and in which they may be kept from settling out by the motion or velocity of the water in which they are suspended

Symbiont: one of the organisms in a symbiotic relationship

Symbiosis: the relationship of two or more different organisms in a close association that may be but is not necessarily of benefit to each

Symbiotic: said of the relationship of two or more different organisms in a close association that may be but is not necessarily of benefit to each

Tannin (tannic acid): any of various chemically different substances capable of promoting the tanning of leather

Terrace: a raised bank of earth that has vertical or sloping sides and a flat top; a flat, narrow stretch of ground that has a steep slope facing a river of sea

Terrestrial: associated with the land

Third-order stream: *see* Stream order

Tidal flat: a former floodplain of a drowned river that is covered during high tide and exposed during low tide

Topography: the features of a small place of region

Trunk: *see* Stem

Upwelling: the movement of cold water up from the bottom to mix with the warmer water at the surface

Vascular plant: any plant that contains vessels

Vertebrate: an animal with a backbone

Water catchment: a drainage basin that catches water—either rain, snow, or both—and stores it in a slow-motion downward flow as it merges with ever-larger water catchments until it is finally accepted into the sea

Water column: the vertical space of water that exists between the bottom and the surface of a body of water, such as a stream or lake

Water-logged: soaked or saturated with water

Water table: the surface in a permeable body of rock of a zone saturated with water

Xylophagid: the vernacular for the generic name *Xylophaga*

Zooplankton: animal organisms, generally microscopic, that float or drift in great numbers in fresh or salt water

ENDNOTES

1. Hall, Martin, Marco Garcia, Cleridy Lennert, and Pablo Arenas. 1992. The Association of Tunas with Floating Objects and Dolphins in the Eastern Pacific Ocean. III. Characteristics of Floating Objects and Their Attractiveness for Tunas, paper given at the Inter-American Tropical Tuna Commission, La Jolla, Calif., 33 pp.
2. Xavier, Ariz, Delgado Alicia, Fonteneau Alain, Gonzales Costas, Fernando Pilar, and Pallares Pilar. 1992. Logs and Tunas in the Eastern Tropical Atlantic, A Review of Present Knowledge and Uncertainties, paper given at the Inter-American Tropical Tuna Commission, La Jolla, Calif., 23 pp.
3. Anonymous. 1976. *Atlas Climatológico e Hidrográfico del Istom Centroamericano,* Publ. Instituto Panamericano de Geografía e Historia, 367 pp.
4. Hoffmann, J.A.J. 1975. *Climatic Atlas of South America. I. Maps of Mean Temperature and Precipitation,* OMM.WMO, UNESCO, Paris, and Cartographia, Budapest.
5. Steinhauser, F. 1979. *Climatic Atlas of North and Central America. I. Maps of Mean Temperature and Precipitation,* OMM.WMO, UNESCO, Paris, and Cartographia, Budapest.
6. Nials, Fred L., Eric E. Deeds, Michael E. Mosley, Shelia G. Pozorski, Thomas G. Pozorski, and Robert Feldman. 1979. El Niño: The catastrophic flooding of coastal Peru. *Field Museum Nat. Hist. Bull.* 50(7):4–14
7. Nials, Fred L., Eric E. Deeds, Michael E. Mosley, Shelia G. Pozorski, Thomas G. Pozorski, and Robert Feldman. 1979. El Niño: The catastrophic flooding of coastal Peru. *Field Museum Nat. Hist. Bull.* 50(8):4–10.
8. Quinn, William H., Victor T. Neal, and Santiago E. Antunez de Mayolo. 1987. El Niño occurrences over the past four and a half centuries. *J. Geophys. Res.* 92:14,449–14,461.
9. Xeres, F. *Verdadera Relacion de la Conquista del Peru, Seville, 1534* (Reports on the Discovery of Peru), translated and edited by C.R. Markham, Burt Franklin, New York, 1872.
10. Oliva, Anello. 1631. *Historia del Peru y Varones Insignes en Santidad de la Compania de Jesus,* reprinted by Juan Francisco Pazos Varela and Luis

Varela y Orbegozo (Eds.), Lima Imp. y Liv. de San Pedro, Lima, 1895, 216 pp.

11. Alongi, D. 1990. The ecology of tropical soft-bottom benthic ecosystems. *Oceanogr. Mar. Biol. Annu. Rev.* 23:381–496.

12. Biswas, A.K. (Ed.). 1978. Register of international rivers. *Water Supply Manage.* 2:1–58.

13. Hatcher, B.G., R.E. Johannes, and A.I. Robertson. 1989. Review of research relevant to the conservation of shallow tropical marine systems. *Oceanogr. Mar. Biol. Annu. Rev.* 27:337–414.

14. Lyell, Sir Charles. 1866. *Principles of Geology; or The Modern Changes of the Earth and Its Inhabitants*, D. Appleton, New York, 834 pp.

15. Suzuki, Ziro. 1992. General Description on Tuna Biology Related to Fishing Activities on Floating Objects by Japanese Purse Seine Boats in the Western and Central Pacific, paper given at the Inter-American Tropical Tuna Commission, La Jolla, Calif., 9 pp.

16. Swan, James G. 1857. *The Northwest Coast*, University of Washington Press, Seattle, 435 pp. (reprinted in 1982).

17. Secretary of the Treasury. 1859. The report of the superintendent of the coast survey showing the progress of the survey in 1858. in House Executive Documents, No. 33: 2nd session, 35th Congress, U.S. Government Printing Office, Washington, D.C.

18. Swan, James G. 1971. *Almost Out of This World. Scenes from Washington Territory*, Washington State Historical Society, Tacoma, 126 pp.

19. Report of the Secretary of War. 1875–1921. Report of the Chief of Engineers. in House Executive Documents, Sessions of Congress, Annual Reports, U.S. Government Printing Office, Washington, D.C.

20. Report of the Secretary of War. 1883–84. Report of the Chief of Engineers, Vol. 2, Part 1, in House Executive Documents, Vol. 5, 2nd session, 48th Congress, U.S. Government Printing Office, Washington, D.C.

21. Oregon Administrative Rules, Chapter 736, Division 2, State Parks and Recreation Division, Beach Log and Driftwood Removal Policy, April 1985.

22. Washington State's Water Drift Logs: Hazard or Boon? A 1978 Report, pp. 20–23.

23. Ogden, Peter Skene. 1961. Peter Skene Ogden's snake country journal 1826–27. *Hudson's Bay Rec. Soc.* 23:1–122.

24. Williams, I.A. 1914. Drainage of farm lands in the Willamette and tributary valleys of Oregon. The mineral resources of Oregon. *Ore. Bur. Mines and Geol.* 1:13.

25. Report of the Secretary of War. 1904–5. Report of the Chief of Engineers. Vol. 7, Part 3. in House Executive Documents, Vol. 8, 1st session, 59th Congress, U.S. Government Printing Office, Washington, D.C.

26. Dodge, Orville. 1898. *Pioneer History of Coos and Curry Counties*, Coos-

Curry Pioneer and Historical Association, Bandon, Ore., 468 pp. (reprinted in 1969).

27. Fries, Robert F. 1948. The Mississippi River Logging Company and the struggle for the free navigation of logs, 1865–1900. *Mississippi Valley Hist. Rev.* 35:429–448.

28. Morgan, Joe A. 1942. When the Chippewa Forks were driving streams. *Wisc. Mag. Hist.* 26:391–407.

29. Rector, William G. 1949. From woods to sawmill: Transportation problems in logging. *Agric. Hist.* 23:239–244.

30. Sedell, James R., Data on file at the Pacific Northwest Research Station, Forestry Sciences Laboratory, 3200 Jefferson Way, Corvallis, OR 97331.

31. Sedell, James R., Peter A. Bisson, Fredrick J. Swanson, and Stanley V. Gregory. 1988. What we know about large trees that fall into streams and rivers. in From the Forest to the Sea—A Story of Fallen Trees, Chris Maser, Robert F. Tarrant, James M. Trappe, and Jerry F. Franklin (Tech. Eds.), USDA Forest Service General Technical Report PNW-229, Pacific Northwest Research Station, Portland, Ore.

32. Baker, C.O. 1979. The Impacts of Logjam Removal on Fish Populations and Stream Habitat in Western Oregon, Ph.D. thesis, Oregon State University, Corvallis, 86 pp.

REFERENCES

Albion, R.G. 1926. Forests and sea power: The timber problem of the Royal Navy. in *Harvard Economic Studies*, Harvard University Press, Cambridge, Mass., pp. 1652–1865.

Allee, B.A. 1982. The role of interspecific competition in the distribution of salmonids in streams. in *Proceedings of the Salmon and Trout Migratory Behavior Symposium*, University of Washington, Seattle, pp. 111–122.

Anderson, N.H. 1989. Xylophagous Chironomidae from Oregon streams. *Aquatic Insects* 11:33–45.

Anderson, N.H. and J.R. Sedell. 1979. Detritus processing by macroinvertebrates in stream ecosystems. *Annu. Rev. Entomol.* 24:351–377.

Anderson, N.H., J.R. Sedell, L.M. Roberts, and F.J. Triska. 1978. The role of aquatic invertebrates in processing wood debris from coniferous forest streams. *Am. Midl. Nat.* 100:64–82.

Anderson, N.H., R.J. Steedman, and T. Dudley. 1984. Patterns of exploitation by stream invertebrates of wood debris (xylophagy). *Verh. Internatl. Verein. Limnol.* 22:1847–1852.

Andersen, N.M. and J.T. Polhemus. 1976. Water-striders (Hemiptera:Gerridae, Veliidae, etc.) in *Marine Insects*, L. Cheng (Ed.), American Elsevier, New York, pp. 187–224.

Anonymous. 1989. Inter-American Tropical Tuna Commission, Dolphin Work-shop held at San Jose, Costa Rica, March 14–16, 1989 (working documents 1–4).

Bachman, R.A. 1984. Foraging behavior of free-ranging wild and hatchery brown trout in a stream. *Trans. Am. Fish. Soc.* 113:1–32.

Bailey, V. 1936. The mammals and life zones of Oregon. *N. Am. Fauna* 55:1–416.

Baker, J.H., R.Y. Morita, and N.H. Anderson. 1983. Bacterial activity associated with the decomposition of wood substrates in a stream sediment. *J. Appl. Environ. Microbiol.* 45:516–521.

Barnes, C.A., A.C. Duxbury, and B.A. Morse. 1971. Circulation and selection properties of the Columbia River effluent at sea. in *The Columbia River Estuary and Adjacent Waters Bioenvironmental Studies*, A.T. Pruter and D.L. Alverson (Eds.), University of Washington Press, Seattle, pp. 41–80.

Bayer, R.D. 1978. Aspects of Oregon estuarine great blue heron population. in *Wading Birds*, A. Sprunt, IV, J.C. Ogden, and S. Winckler (Eds.), National Audubon Society Research Paper 7, pp. 213–217.

Bayer, R.D. 1980. Birds feeding on herring eggs at the Yaquina estuary, Oregon. *Condor* 82:193–198.

Bayer, R.D. 1981. California sea lions in the Yaquina River estuary, Oregon. *Murrelet* 62:56–59.

Bayer, R.D. 1983. Nesting success of western gulls at Yaquina Head and on man-made structures in Yaquina estuary, Oregon. *Murrelet* 64:87–91.

Bayer, R.D. 1985. Six years of harbor seal censusing at Yaquina estuary, Oregon. *Murrelet* 66:44–49.

Begley, S., L. Drew, and M. Hager. 1989. Smothering the waters. *Newsweek* April 10:54–57.

Bell, J.D. and D.A. Pollard. 1989. Ecology of fish assemblages and fisheries associated with seagrasses. in *Biology of Seagrasses*, A.W. Larkun, A.J. McComb, and S.A. Sheperd (Eds.), Elsevier, Amsterdam, pp. 565–609.

Benke, A.C. and J.B. Wallace. 1990. Wood debris in coastal plain blackwater streams. *Can. J. Fish. Aquat. Sci.* 47:92–99.

Bent, A.C. 1964. *Life Histories of North American Nuthatches, Wrens, Thrashers, and Their Allies*, Dover, New York, 475 pp.

Beschta, R.L. 1983. The effects of large organic debris upon channel morphology: A flume study. in *Proceedings, D.B. Simons Symposium on Erosion and Sedimentation*, Simons, Li, and Assoc., Fort Collins, Colo., pp. 8.63–8.78.

Besednov, L.N. 1960. Some data on the ichthyofauna of Pacific Ocean flotsam. *Trudy Inst. Okeanol.* 41:192–197.

Bigg, M.A. 1969. Clines in the pupping season of the harbor seal, *Phoca vitulina*. *Fish. Res. Board Can. J.* 26:449–455.

Bigg, M.A. 1969. The harbour seal in British Columbia. *Fish. Res. Board Can. Bull.* 172:1–33.

Bilby, R.E. 1981. Role of organic dams in regulating the export of dissolved and particulate matter from a forested watershed. *Ecology* 62:1234–1243.

Bilby, R.E. 1984. Post-logging removal of woody debris affects stream channel stability. *J. For.* 82:609–613.

Bilby, R.E. 1985. Influence of stream size on the function and characteristics of large organic debris. in *Proceedings West Coast Meeting of National Council of Air and Stream Improvement*, National Council Pap. Indust. Air and Stream Improvement, Portland, Ore., pp. 1–14.

Bisson, P.A., J.L. Nielson, R.A. Palmason, and L.E. Grove. 1982. A system of naming habitat types in small streams, with examples of habitat utilization by salmonids during low streamflow. in *Acquisition and Utilization of Aquatic Habitat Information*, N.B. Armantrout (Ed.), Western Division of the American Fisheries Society, pp. 62–73.

Bisson, P.A., J.L. Nielson, and J.W. Ward. 1985. Experiment release of coho salmon (*Oncorhynchus kisutch*) into a stream impacted by Mount Saint Helens volcano. *Proc. West. Assoc. Fish. Wildl. Agen.* 1984:422–435.

Bisson, P.A. and J.R. Sedell. 1984. Salmonid populations in streams in clearcut vs. old-growth forests of western Washington. in *Fish and Wildlife Relationships in Old-Growth Forests*, W.R. Meehan, T.R. Merrell, Jr., and T.A. Hanley (Eds.), American Institute of Fisheries Research Biology, Juneau, Alaska, pp. 121–129.

Bisson, P.A., R.E. Bilby, M.D. Bryant, et al. 1987. Large woody debris in forested streams in the Pacific Northwest: Past, present, and future. in *Proceedings of a Symposium; Streamside Management–Forestry and Fisheries Interactions*, T. Cundy and E. Salo (Eds.), University of Washington, Seattle, pp. 143–190.

Boehlert, G.W. and A. Genin. 1987. A review of the effects of seamounts on biological processes. in *Seamounts, Islands, and Atolls*, B.H. Keating, P. Fryer, R. Batiza, and G.W. Boehlert (Eds.), Geophysical Monograph 43, pp. 319–334.

Bourke, R.H., B. Glenne, and B.W. Adams. 1971. Environment of the Pacific Northwest Coast, Ref. 71-45, Department of Oceanography, Oregon State University, Corvallis, 127 pp.

Brill, R.W., K.N. Holland, and J.S. Ferguson. 1984. Use of ultrasonic telemetry to determine the short-term movements and residence times of tunas around fish aggregating devices. in *Proceedings of the Pacific Congress of Maritime Technology PACON*, Marine Research Management, Honolulu, pp. 1–7.

Brock, R.E. 1985. Fish aggregation devices: How they work and their place in fisheries enhancement. in *Proceedings of the 1st World Angling Conference,* R.H. Strond (Ed.), International Game and Fish Association, Fort Lauderdale, Fla., pp. 193–202.

Brunn, A.F. 1957. Deep sea and abyssal depths. in *Treatise on Marine Ecology and Paleoecology,* Memoir 67, Vol. 1: Ecology, J.W. Hedgpeth (Ed.), Geological Society of America, pp. 641–672.

Bryant, M.D. 1980. Evolution of Large, Organic Debris after Timber Harvest: Maybeso Creek, 1949 to 1978, USDA Forest Service General Technical Report PNW-101, Pacific Northwest Forestry and Range Experiment Station, Portland, Ore., 30 pp.

Bryant, M.D. 1983. The role and management of woody debris in west coast salmonid nursery streams. *N. Am. J. Fish. Manage.* 3:322–330.

Bryant, M.D. 1985. Changes 30 years after logging in large woody debris and its use by salmonids. in *Riparian Ecosystems and Their Management: Reconciling Conflicting Uses,* Proceedings of the 1st North American Riparian Conference, R.R. Johnson, C.D. Ziebel, B.R. Patton, P.F. Ffolliott, and R.H. Hamre (Eds.), USDA Forest Service General Technical Report RM-120, Rocky Mountains Forestry and Range Experiment Station, Fort Collins, Colo., pp. 329–334.

Buckley, B.M. and F.J Triska. 1978. Presence and ecological role of nitrogen-fixing bacteria associated with wood decay in streams. *Internatl. Verein. Theor. Angew. Limnol. Verhandl.* 20:1333–1339.

Bustard, D.R. and D.W. Narver 1975. Aspects of the winter ecology of juvenile coho salmon (*Oncorhynchus kisutch*) and steelhead trout (*Salmo gairdneri*). *J. Fish. Res. Board Can.* 32:667–680.

Bustard, D.R. and D.W. Narver. 1975. Preferences of juvenile coho salmon (*Oncorhynchus kisutch*) and cutthroat trout (*Salmo clarki*) relative to simulated alteration of winter habitat. *J. Fish. Res. Board Can.* 32:681–687.

Caddy, John F. and Jacek Majkowski. 1992. Trees and Tuna: A Reflection on the Long-Term Perspectives for Tuna Fishing around Floating Timber, paper given at the Inter-American Tropical Tuna Commission, La Jolla, Calif., 6 pp.

Center for Marine Conservation. 1988. *A Citizen's Guide to Plastics in the Ocean: More than a Litter Problem,* Center for Marine Conservation, Washington, D.C., 143 pp.

Chaney, R.W. 1956. *The Ancient Forests of Oregon,* Condon Lectures, University of Oregon Press, Eugene, 56 pp.

Cheng, L. 1972. Skaters of the seas. *Oceans* 5:54–55.

Chittenden, Hiram Martin. 1962. *History of Early Steamboat Navigation on the Missouri River*, Vol. 1, Ross & Haines, Minneapolis, 461 pp.

Cole, J.S. 1980. Synopsis of biological data on the yellowfin tuna, *Thunnus albacares* (Bonnaterre, 1788), in the Pacific Ocean, *Inter-American Tropical Tuna Commission Special Report* 2:71–212.

Cottingham, D. 1988. Persistent Marine Debris, Challenge and Response, The Federal Perspective, Grant No. NA86AA-D-SG041, Project No. A/75-01, NOAA Office of Sea Grant and Extramural Programs, U.S. Department of Commerce, 41 pp.

Cummins, K.W. and M.A. Wilzbach. 1985. *Field Procedures for Analysis of Functional Feeding Groups of Stream Macroinvertebrates*, Contrib. 1611, Appalachian Environmental Laboratory, University of Maryland, Frostburg, 18 pp.

Damant, G.C.C. 1921. Illumination of plankton. *Nature* 108:42–43.

Davies, B.R. and K.F. Walker (Eds.). 1986. *The Ecology of River Systems, Monographae Biologicae*, DR W. Junk, Dordrecht, The Netherlands.

Dayton, P.K. 1971. Competition, disturbance, and community organization: The provision and subsequent utilization of space in a rocky intertidal community. *Ecol. Monogr.* 41:351–389.

Dean, R.C. 1976. Cellulose and wood digestion in the marine mollusc *Bankia gouldi* Bartsch. in *International Biodegradation Symposium*, Applied Science Press, London, pp. 955–965.

Dill, L.M., R.C. Ydenberg, and H.G. Fraser. 1981. Food abundance and territory size in juvenile coho salmon (*Oncorhynchus kisutch*). *Can. J. Zool.* 59:1801–1809.

Dolloff, C.A. 1983. The Relationships of Wood Debris to Juvenile Salmonid Production and Microhabitat Selection in Small Southeast Alaska Streams, Ph.D. thesis, Montana State University, Bozeman, 100 pp.

Dudley, T.L. 1982. Population and Production Ecology of *Lipsothrix* spp. (Diptera:Tipulidae), M.S. thesis, Oregon State University, Corvallis, 161 pp.

Dudley, T. and N.H. Anderson. 1982. A survey of invertebrates associated with wood debris in aquatic habitats. *Melanderia* 39:1–21.

Dudley, T. and N.H. Anderson. 1987. The biology and life cycles of *Lipsothrix* spp. (Diptera:Tipulidae) inhabiting wood in Western Oregon streams. *Freshwater Biol.* 17:437–451.

Eilers, H.P. 1975. Plants, Plant Communities, Net Production and Tide Levels: The Ecological Biogeography of the Nehalem Salt Marshes, Tillamook County, Oregon, Ph.D. thesis, Oregon State University, Corvallis, 368 pp.

Elliott, S. and D. Hubartt. 1979. Study of Land Use Activities and Their Relationship to Sport Fish Resources in Alaska: Sport Fish in Alaska, Annual Performance Report, July 1, 1977–June 30, 1978, Vol. 19, Federal Aid in Fish Restoration, Department of Fisheries and Game, Juneau, Alaska, 52 pp.

Eltringham, S.K. 1971. Marine borers and fungi. in *Marine Borers, Fungi and Fouling Organisms of Wood*, E.B.G. Jones and S.K. Eltringham (Eds.), Organ. Econ. Coop. and Devel., Paris, pp. 327–335.

Fausch, K.D. 1984. Profitable stream positions for salmonids: Relating specific growth rate to net energy gain. *Can. J. Zool.* 62:441–451.

Fedoryako, B.I. 1982. Langmuir circulations and a possible mechanism of formation of fish associations around a floating object. *Oceanology* 22:228–232.

Forsbergh, E.D. 1980. Synopsis of biological data on the skipjack tuna *Katsuwonus pelamis* (Linnaeus, 1758), in the Pacific Ocean. *Inter-American Tropical Tuna Commission Special Report* 2:295–360.

Franklin, J.F. and C.T. Dyrness. 1973. Natural Vegetation of Oregon and Washington, USDA Forest Service General Technical Report PNW-8, Pacific Northwest Forestry and Range Experiment Station, Portland, Ore., 417 pp.

Franklin, J.F., K. Cromack, Jr., W. Denison, A. McKee, C. Maser, J. Sedell, F. Swanson, and G. Juday. 1981. Ecological Characteristics of Old-Growth Douglas-Fir Forests, USDA Forest Service General Technical Report PNW-118, Pacific Northwest Forestry and Range Experiment Station, Portland, Ore., 48 pp.

Fries, N. 1966. Chemical factors in the germination of spores of Basidiomycetes. in *The Fungus Spore*, M.F. Madelin (Ed.), Butterworths, London, pp. 189–199.

Fries, N. 1982. Effects of plant roots and growing mycelia on Basidiospore germination in mycorrhiza-forming fungi. in *Arctic and Alpine Mycology*, G.A. Laursen and J.F. Ammirati (Eds.), University of Washington Press, Seattle, pp. 493–508.

Froehlich, H.A. 1973. Natural and man-caused slash in headwater streams. in *Loggers Handbook*, Vol. 33, Pacific Logging Congress, Portland, Ore., 8 pp.

Gabrielson, I.N. and S.G. Jewett. 1970. *Birds of the Pacific Northwest, with Special Reference to Oregon*, Dover, New York, 650 pp.

Gharrett, J.T. and J.I. Hodges. 1950. *Salmon Fisheries of the Coastal Rivers of Oregon South of the Columbia*, Contrib. No. 13, Oregon Fish Commission, Portland, 31 pp.

Gibson, R.J. 1981. Behavioural Interactions between Coho Salmon (*Oncorhynchus kisutch*), Atlantic Salmon (*Salmo salar*), Brook Trout (*Salvenlinus fontinalis*), and Steelhead Trout (*Salmo gairdneri*) at the Juvenile Fluviatile Stages, Can. Tech. Rep. 1029, Fisheries and Aquatic Science, Research and Resource Service, St. John's, Newfoundland.

Gonor, J.J., J.R. Sedell, and P.A. Benner. 1988. What we know about large trees in estuaries, in the sea, and on coastal beaches. in From the Forest to the Sea, A Story of Fallen Trees, C. Maser, R.F. Tarrant, J.M. Trappe, and J.F. Franklin (Tech. Eds.), USDA Forest Service General Technical Report GTR-PNW-229, Pacific Northwest Research Station, Portland, Ore., pp. 83–112.

Gooding, R.M. and J.J. Magnuson. 1967. Ecological significance of a drifting object to pelagic fishes. *Pac. Sci.* 21:486–497.

Gosselink, J.L., G.P. Shaffer, L.C. Lee, et al. 1990. Landscape conservation in a forested wetland watershed. *BioScience* 40:588–600.

Grassle, J.F. 1986. Hydrothermal vent animals: Distribution and biology. *Science* 229:713–717.

Greenblatt, P.R. 1979. Associations of tuna with flotsam in the eastern tropical Pacific. *Fish. Bull.* 77:147–155.

Gregory, S.V., F.J. Swanson, W.A. McKee, and K.W. Cummins. 1991. An ecosystem perspective of riparian zones. *BioScience* 41:540–551.

Grette, G.B. 1985. The Abundance and Role of Large Organic Debris in Juvenile Salmonid Habitat Streams in Second Growth and Unlogged Forests, M.S. thesis, University of Washington, Seattle, 105 pp.

Hall, J.D. and C.O. Baker. 1982. Rehabilitating and enhancing stream habitat. I. Review and evaluation. in Influence of Forest and Rangeland Management on Anadromous Fish Habitat in Western North America, W.R. Meehan (Ed.), USDA Forest Service General Technical Report PNW-138, Pacific Northwest Forestry and Range Experiment Station, Portland, Ore., pp. 1–29.

Hall, Martin, Pablo Arenas, and Forest Miller. 1992. The Association of Tunas with Floating Objects and Dolphins in the Eastern Pacific Ocean. I. Environment and Fishing Areas, paper given at the Inter-American Tropical Tuna Commission, La Jolla, Calif., 26 pp.

Hall, Martin, Marco Garcia, Cleridy Lennert, and Pablo Arenas. 1992. The Association of Tunas with Floating Objects and Dolphins in the Eastern Pacific Ocean. III. Characteristics of Floating Objects and Their Attractiveness for Tunas, paper given at the Inter-American Tropical Tuna Commission, La Jolla, Calif., 33 pp.

Hall, Martin, Marco Garcia, Alejandro Pares-Sierra, and Pablo Arenas. 1992. The Association of Tunas with Floating Objects and Dolphins in the Eastern Pacific Ocean. V. Simulated Trajectories of Floating Objects Entering the Eastern Pacific Ocean, paper given at the Inter-American Tropical Tuna Commission, La Jolla, Calif., 8 pp.

Harmon, M.E., J.F. Franklin, F.J. Swanson, P. Sollins, S.V. Gregory, J.D. Lattin, N.H. Anderson, S.P. Cline, N.G. Sumen, J.R. Sedell, G.W. Lienkaemper, K. Cromack, Jr., and K.W. Cummins, 1986. Ecology of coarse woody debris in temperate ecosystems. *Adv. Ecol. Res.* 15:133–302.

Harris, L.D. 1984. *The Fragmented Forest,* University of Chicago Press, Chicago, 211 pp.

Harris, S.L. 1976. *Fire and Ice,* Pacific Search Press, Seattle, 320 pp.

Hedges, J.I., J.R. Ertel, and E.B. Leopold. Lignin geochemistry of a late quaternary sediment core from Lake Washington. *Geochim. Cosmochim. Acta* 46:1869–1877.

Hedges, J.I. and D.C. Mann. 1979. The lignin geochemistry of marine sediments from the southern Washington coast. *Geochim. Cosmochim. Acta* 43:1809–1818.

Heede, B.H. 1972. Influences of a forest on the hydraulic geometry of two mountain streams. *Water Resour. Bull.* 8:523–530.

Heede, B.H. 1985. The evolution of salmonid streams. in *Proceedings of the Symposium Wild Trout III,* F. Richardson and R.H. Hamre (Eds.), Trout Unlimited, Vienna, Va., pp. 33–37.

Hogan, D. 1985. The influence of large organic debris on channel morphology in Queen Charlotte Island streams. *Proc. West. Assoc. Fish. Wildl. Agen.* 1984:263–273.

House, R.A. and P.L. Boehne. 1985. Evaluation of instream enhancement structures for salmonid spawning and rearing in a coastal Oregon stream. *N. Am. J. Fish. Manage.* 5:283–295.

Hunter, J.R. and C.T. Mitchell. 1968. Field experiments on the attraction of pelagic fish to floating objects. *J. Cons.* 31:427–434.

Hynes, H.B.N. 1972. *The Ecology of Running Waters,* University of Toronto Press, Toronto, 555 pp.

Inone, M., R. Amano, Y. Iwasaki, and M. Yamanati. 1968. Studies on environments alluring skipjack and other tunas. 2. On the driftwoods accompanied by skipjack and tunas. *Bull. Jap. Soc. Sci. Fish.* 34:283–287.

Jannasch, H.W. and M.J. Mottl. 1985. Geomicrobiology of deep-sea hydrothermal vents. *Science* 229:717–725.

Johannessen, C.L. 1964. Marshes prograding in Oregon: Aerial photographs. *Science* 146:1575–1578.

Johnson, G.E. and J.J. Gonor. 1982. The tidal exchange of *Callianassa californiensis* larvae between the ocean and the Salmon River estuary. *Estuar. Coastal Shelf Sci.* 14:511–516.

Johnson, M.W. 1935. Seasonal migration of the wood-borer *Limnoria lignorum* at Friday Harbor, Washington. *Biol. Bull.* 69:427–438.

Jones, E.B.G., H. Kuhne, P.C. Trussell, and R.D. Turner. 1972. Results of an international cooperative research programme on the biodeterioration of timber submerged in the sea. *Material Organismen.* 7:93–118.

Jones, E.B.G., R. Turner, S.E. Furtado, and H. Kuhne. 1976. Marine biodeteriogenic organisms. I. Lignicolous fungi and bacteria and wood boring mollusca and crustacea. *Int. Biodeterior. Bull.* 12:120–134.

Kaufmann, P.R. 1987. Channel Morphology and Hydraulic Characteristics of Torrent-Impacted Forest Streams in the Oregon Coast Range, USA, Ph.D dissertation, Oregon State University, Corvallis, 235 pp.

Keller E.A. and F.J. Swanson. 1979. Effects of large organic material on channel form and fluvial processes. *Earth Surf. Process.* 4:361–380.

Keller, E.A. and T. Tally. 1979. Effects of large organic debris on channel form and fluvial processes in the coastal redwood environment. in *Adjustments on the Fluvial System*, Proc. 10th Annual Geomorphol. Symposium, D.D. Rhodes and G.P. Williams (Eds.), State University of New York, Binghamton, pp. 169–197.

Kerst, C.D. 1970. The Seasonal Occurrence and Distribution of Stoneflies (Plecoptera) of a Western Oregon Stream, M.S. thesis, Oregon State University, Corvallis, 80 pp.

Klimley, A.P. and S.B. Butler. 1988. Immigration and emigration of pelagic fish assemblages to seamounts in the Gulf of California related to water mass movements using satellite imagery. *Mar. Ecol. Progr. Ser.* 49:11–22.

Klimley, A.P., S.B. Butler, A. Stull, and D.R. Nelson. 1988. Diurnal movements of scalloped hammerhead sharks (*Sphyrna lewini* Griffith and Smith) to and from a seamount in the Gulf of California. *J. Fish. Biol.* 33:751–761.

Kodata, H. 1958. Cellulose-decomposing bacteria in the sea. in *Marine Wood Boring and Fouling Organisms*, D.L. Ray (Ed.), University of Washington Press, Seattle, pp. 332–341.

Kojima, S. 1960. Studies of dolphin fishing conditions in the western Sea of Japan.

V. On the species of fishes attracted to bamboo rafts. *Bull. Jap. Soc. Sci. Fish.* 26:379–382.

Kojima, S. 1960. Studies of dolphin fishing conditions in the western Sea of Japan. VI. On the ecology of the groups of fish congregating around bamboo rafts. *Bull. Jap. Soc. Sci. Fish.* 26:383–388.

Komar, P.D. 1983. The erosion of Siletz Spit, Oregon. in *Handbook of Coastal Processes and Erosion,* P.D. Komar (Ed.), CRC Press, Boca Raton, Fla.

Komar, P.D. and C.C. Rea. 1976. Erosion of Siletz Spit, Oregon. *Shore and Beach* 44:9–15, 65–77.

Lestelle, L.C. 1978. The Effects of Forest Debris Removal on a Population of Resident Cutthroat Trout in a Small Headwater Stream, M.S. thesis, University of Washington, Seattle, 133 pp.

Lestelle, L.C. and C.J. Cederholm. 1984. Short-term effects of organic debris removal on resident cutthroat trout. in *Fish and Wildlife Relationships in Old-Growth Forests,* W.R. Meehan, T.R. Merrell, Jr., and T.A. Hanley (Eds.), American Institute of Fisheries Research Biology, Juneau, Alaska, pp. 131–140.

Levings, C.D., L.B. Holtby, and M.A. Henderson (Eds.). 1989. Proceedings of the National Workshop on Effects of Habitat Alteration on Salmonid Stocks, Canadian Special Publication of Fisheries and Aquatic Sciences 105, Department of Fisheries and Oceans, Ottawa, 199 pp.

Li, C.Y. and M.A. Castellano. 1985. Nitrogen-fixing bacteria isolated from within sporocarps of three ectomycorrhizal fungi. in *Proceedings 6th North American Conference on Mycorrhiza,* R. Molina (Ed.), Forestry Research Laboratory, Corvallis, Ore., p. 164.

Li, C.Y., C. Maser, Z. Maser, and B.A. Caldwell. 1986. Role of three rodents in forest nitrogen fixation in western Oregon: Another aspect of mammal-mycorrhizal fungus-tree mutualism. *Great Basin Nat.* 46:411–414.

Lisle, T.E. 1982. Roughness elements: A key resource to improve anadromous fish habitat. in *Proceedings, Propagation, Enhancement, and Rehabilitation of Anadromous Salmonid Populations and Habitat in the Pacific Northwest Symposium,* Cooperative Fisheries Research Unit, Humboldt State University, Arcata, Calif., pp. 93–98.

Lisle, T.E. and H.M. Kelsey. 1982. Effects of large roughness elements on the thalweg course and pool spacing. in *American Geomorphological Field Group Field Trip Guidebook,* American Geophysics Union, Berkeley, Calif., pp. 134–135.

Lister, D.B. and H.S. Genoe. 1979. Stream habitat utilization by cohabiting underyearlings of chinook (*Oncorhynchus tschawytscha*) and coho salmon (*O. kisutch*) in the Big Qualicum River, British Columbia. *J. Fish. Res. Board Can.* 27:1215–1224.

Lowe-McConnell, R.H. 1975. *Fish Communities in Tropical Freshwater: Their Distribution, Ecology and Evolution*, Longman, New York.

Luternauer, J.L., J.J. Clague, K.W. Conway, J.V. Barrie, B. Blaise, and R.W. Mathewes. 1989. Late Pleistocene terrestrial deposits on the continental shelf of western Canada: Evidence for rapid sea-level change at the end of the last glaciation. *Geology* 17:357–360.

Macdonald, J.S., I.K. Birtwell, and G.M. Kruzynski. 1987. Food and habitat utilization by juvenile salmonids in the Campbell River estuary. *Can. J. Fish. Aquat. Sci.* 44:1233–1246.

Macdonald, J.S., C.D. Levings, C.D. McAllister, U.H.M. Fagerlund, and J.R. McBride. 1988. A field experiment to test the importance of estuaries for chinook salmon (*Oncorhynchus tschawytscha*) survival: Short-term results. *Can. J. Fish. Aquat. Sci.* 45:1366–1377.

Marzolf, G.R. 1978. The Potential Effects of Clearing and Snagging on Stream Ecosystems, FWS/OBS-78/14:1–31, Biological Services Program, Fish and Wildlife Service, U.S. Department of the Interior.

Maser, C. 1973. Preliminary notes on the distribution, ecology, and behavior of *Cicindela bellissima* Leng. *Cicindela* 5:61–76.

Maser, C. 1989. *Forest Primeval, The Natural History of an Ancient Forest*, Sierra Club Books, San Francisco.

Maser, C., B.R. Mate, J.F. Franklin, and C.T. Dyrness. 1981. Natural History of Oregon Coast Mammals, USDA Forest Service General Technical Report PNW-133, Pacific Northwest Forestry and Range Experiment Station, Portland, Ore., 496 pp.

Maser, C. and J.M. Trappe (Tech. Eds.). 1984. The Seen and Unseen World of the Fallen Tree, USDA Forest Service General Technical Report PNW-164, Pacific Northwest Forestry and Range Experiment Station, Portland, Ore., 56 pp.

Maser, C., J.M. Trappe, and R.A. Nussbaum. 1978. Fungal-small mammal interrelationships with emphasis on Oregon coniferous forests. *Ecology* 59:799–809.

Maser, C., R.F. Tarrant, J.M. Trappe, and J.F. Franklin (Tech. Eds.). 1988. From the Forest to the Sea, A Story of Fallen Trees, USDA Forest Service General Technical Report PNW-GTR-229, Pacific Northwest Research Station, Portland, Ore., 153 pp.

Mason, D.T. 1985. Follow-up Study of Habitat Values of the Lower Stehekin River, Summer 1985, Supplementary Report, National Park Service, Contract #CX-9000-3-E066, 16 pp.

Mason, D.T. and J. Koon. 1958. Habitat Values of Woody Debris Accumulations of the Lower Stehekin River, with Notes on Disturbances to Alluvial Gravels, Final Report, National Park Service, Contract #CX-9000-3-E066, 140 pp.

McKernan, D.L., D.R. Johnson, and J.I. Hodges. 1950. Some factors influencing the trends of salmon populations in Oregon. *Trans. N. Am. Wildl. Conf.* 15:427–449.

McMahon, Thomas E. and Gordon F. Hartman. 1989. Influence of cover complexity and current velocity on winter habitat use by juvenile coho salmon (*Oncorhynchus kisutch*). *Can. J. Fish. Aquat. Sci.* 46:1551–1557.

McMahon, Thomas E. and L. Blair Holtby. 1992. Behaviour, habitat use, and movements of coho salmon (*Oncorhynchus kisutch*) smolts during seaward migration. *Can. J. Fish. Aquat. Sci.* 49:1478–1485.

McNeely, R.L. 1961. Purse seine revolution in tuna fishing. *Pac. Fisher.* 59:27–58.

Meehan, W.R., W.A. Farr, D.M. Bishop, and J.H. Patric. 1969. Some Effects of Clearcutting on Salmon Habitat in Two Southeastern Alaska Streams, USDA Forest Service Research Paper PNW-82, Pacific Northwest Forestry and Range Experiment Station, Portland, Ore., 45 pp.

Merrill, W. and E.B. Cowling. 1966. Role of nitrogen in wood deterioration: Amount and distribution of nitrogen in fungi. *Phytopathology* 56:1083–1090.

Merritt, R.W. and K.W. Cummins (Eds.). 1984. *An Introduction to the Aquatic Insects of North America*, 2nd ed., Kendall/Hunt, Dubuque, Iowa, 722 pp.

Minshall, G.W., K.W. Cummins, R.C. Petersen, C.E. Cushing, D.A. Bruns, J.R. Sedell, and R.L. Vannote. 1985. Developments in stream ecosystem theory. *Can J. Fish. Aquat. Sci.* 42:1045–1055.

Morse, E. 1883. *Morse's Monthly: A Puget Sound Magazine for the People of the Northwest. Snohomish City, Washington Territory* 1:1–14 (available from the Washington State Archives, Olympia).

Mundi, J.H. 1969. Ecological implications of the diet of juvenile coho in streams. in *Symposium on Salmon and Trout in Streams*, T.G. Northcote (Ed.), H.R. MacMillan Lectures in Fisheries, Institute of Fisheries, University of British Columbia, Vancouver, pp. 135–152.

Murphy, M.L., J.F. Thedinga, K.V. Koski, and G.B. Grette. 1984. A stream ecosystem in an old-growth forest in southeast Alaska. Part 5. Seasonal changes in habitat utilization by juvenile salmonids. in *Fish and Wildlife Relationships in Old-Growth Forests,* W.R. Meehan, T.R. Merrell, and T.A. Hanley, Jr. (Eds.), American Institute of Fisheries Research Biology, pp. 89–98.

Naiman, R.J. and H. Décamps (Eds.). 1990. *The Ecology and Management of Aquatic–Terrestrial Ecotones,* United Nations Educational, Scientific, and Cultural Organization, Paris, 316 pp.

Naiman, R.J., J.M. Melillo, and J.E. Hobbie. 1986. Ecosystem alteration of boreal forest streams by beaver (*Castor canadensis*). *Ecology* 67:1254–1269.

Narver, D.W. 1971. Effects of logging debris on fish populations. in *Proceedings of the Symposium on Forest Land Uses and Stream Environments,* Oregon State University, Corvallis, pp. 100–111.

National Research Council (U.S.). 1992. *Restoration of Aquatic Ecosystems,* National Academy Press, Washington, D.C.

Newton, R. 1989. Large scale sustainable forestry as a holistic discipline. *Trumpeter* 6:63–65.

Norse, Elliot A. 1993. *Global Marine Biological Diversity, A Strategy for Building Conservation into Decision Making,* Island Press, Washington, D.C., 383 pp.

Nussbaum, R.A., E.D. Brodie, Jr., and R.M. Storm. 1983. *Amphibians and Reptiles of the Pacific Northwest,* Northwest Nature Book, University Press, Moscow, Idaho, 332 pp.

Ogden, P.S. 1961. Peter Skene Ogden's snake country journal 1826–27. *Hudson's Bay Rec. Soc.* 23:1–122.

Oregonian, The. 1986. Logging river banks for firewood after high water idles anglers' sport. March 13.

Osborn, J.G. 1981. The Effects of Logging on Cutthroat Trout (*Salmo clarki*) in Small Headwater Streams, FRI-UW-8113, Fisheries Research Institute, University of Washington, Seattle, 89 pp.

Pacific Tuna Development Foundation. 1979. 1978 Annual Report, Pacific Tuna Development Foundation, Honolulu, 22 pp.

Parin, N.V. 1970. Ichthyofauna of the epipelagic zone. *Nauka Moscow* (Isr. Program Sci. Transl., Jerusalem), 206 pp.

Parin, N.V., V.G. Neiman, and Yu. A. Rudyakov. 1985. To the question of the biological productivity of waters in areas of submerged elevations of the open ocean. in *Biological Basis of the Commercial Exploitation of the Open Areas of the Ocean,* M.E. Vinigradov and M.V. Flint (Eds.), Academy of Sciences of the U.S.S.R., Commission on Problems of the World Ocean, Nauka, Moscow, pp. 192–193.

Parmenter, T. and R. Bailey. 1985. *Oregon Ocean Book,* Department of Conservation and Development and Sea Grant, Oregon State University, Corvallis, 85 pp.

Pearcy, William G. 1992. *Ocean Ecology of North Pacific Salmonids,* University of Washington Press, Seattle, 179 pp.

Pereira, C.R.D. and N.H. Anderson. 1982. Observations on the life histories and feeding of *Cinygma integrum* Eaton and *Ironodes nitidus* (Eaton) (Ephemeroptera: Heptageniidae). *Melanderia* 39:35–45.

Pereira, C.R.D., N.H. Anderson, and T. Dudley. 1982. Gut content analysis of aquatic insects from wood substrates. *Melanderia* 39:23–33.

Perrin, W.F. 1968. The porpoise and the tuna. *Sea Frontiers* 14(3).

Perrin, W.F. 1969. Using porpoise to catch tuna. *World Fish.* 18:42–45.

Petts, G.E., H. Möller, and A.L. Roux. 1989. *Historical Change of Large Alluvial Rivers: Western Europe,* John Wiley & Sons, New York, 355 pp.

Prahl, F.G., J.M. Hayes, and T.-M. Xie. 1992. Diploptene: An indicator of terrigenous organic carbon in Washington coastal sediments. *Limnol. Oceanogr.* 37:1290–1300.

Prahl, F.G., J.R. Ertel, M.A. Goni, M.A. Sparrow, and B. Eversmeyer. (in press). Terrestrial organic carbon contributions to sediments on the Washington margin. *Geochim. Cosmochim. Acta.*

Price, B. 1988. Plastic's threatening tide. *Science World* 44:8–11.

Quayle, D.B. 1956. The British Columbia Shipworm, Report British Columbia Department of Fisheries for 1955, pp. 92–104.

Quinn, William H., Victor T. Neal, and Santiago E. Antunez de Mayolo. 1987. El Niño occurrences over the past four and a half centuries. *J. Geophys. Res.* 92:14,449–14,461.

Rainville, R.P., S.C. Rainville, and E.L. Lider. 1985. Riparian silviculture strategies for fish habitat emphasis. in *Silviculture for Wildlife and Fish: A Time for Leadership,* Proceedings Technical Session Wildlife Fish Ecology Working Group, Society of American Forestry, Bethesda, Md., pp. 186–196.

Ray, D.L. 1959. Nutritional physiology of *Limnoria*. in *Marine Boring and Fouling Organisms*, D.L. Ray (Ed.), University of Washington Press, Seattle, pp. 46–61.

Report of the Secretary of War. 1875–1921. Report of the Chief of Engineers. in House Executive Documents, Sessions of Congress, Annual Reports, U.S. Government Printing Office, Washington, D.C.

Report of the Secretary of War. 1883–84. Report of the Chief of Engineers, Vol. 2, Part 1. in House Executive Documents, Vol. 5, 2nd session, 48th Congress, U.S. Government Printing Office, Washington, D.C.

Report of the Secretary of War. 1894–95. Report of the Chief of Engineers, Vol. 2, Part 1. in House Executive Documents, Vol. 8, 1st session, 54th Congress, U.S. Government Printing Office, Washington, D.C.

Report of the Secretary of War. 1904–5. Report of the Chief of Engineers, Vol. 7, Part 3. in House Executive Documents, Vol. 8, 1st session, 59th Congress, U.S. Government Printing Office, Washington, D.C.

Rhoades, F. 1985. Appendix V, Small animal mycophagy, addendum to Mason, D.T. and J. Koon. 1958. Habitat Values of Woody Debris Accumulations of the Lower Stehekin River, with Notes on Disturbances to Alluvial Gravels, Final Report, National Park Service, Contract #CX-9000-3-E066, 7 pp.

Rhoades, F. 1986. Small mammal mycophagy near woody debris accumulations in the Stehekin River Valley, Washington. *Northw. Sci.* 60:150–153.

Robbins, C.S., B. Bruun, and H.S. Zim. 1983. *A Guide to Field Identification of Birds of North America*, Golden Press, Racine, Wisc., 360 pp.

Rothacher, J., C.T. Dyrness, and R.L. Fredriksen. 1967. Hydrologic and Related Characteristics of Three Small Watersheds in the Oregon Cascades, USDA Forest Service, Pacific Northwest Forestry and Range Experiment Station, Portland, Ore., 54 pp.

Ruth, R.H. and R.A. Yoder. 1953. Reducing Wind Damage in the Forests of the Oregon Coast Range, USDA Forest Service Research Paper No. 7, Pacific Northwest Forestry and Range Experiment Station, Portland, Ore., 30 pp.

Savat, J. 1975. Some morphological and hydraulic characteristics of river-patterns in the Zaire basin. *Catena* 2:161–180.

Scott, J.M. 1969. Tuna schooling terminology. *Calif. Fish Game* 55:136–140.

Secretary of the Treasury. 1859. The report of the superintendent of the coast survey showing the progress of the survey in 1858. in House Executive Documents, No. 33: 2nd session, 35th Congress, U.S. Government Printing Office, Washington, D.C.

Sedell, J.R. and F.J. Swanson. 1984. Ecological characteristics of streams in old-growth forests of the Pacific Northwest. *in Fish and Wildlife Relationships in Old-Growth Forests*, W.R. Meehan, T.R. Merrell, Jr., and T.A. Hanley (Eds.), American Institute of Fisheries Research Biology, Juneau, Alaska, pp. 9–16.

Sedell, J.R., F.J. Swanson, and S.V. Gregory. 1984. Evaluating fish response to woody debris. *in Proceedings of the Pacific Northwest Stream Habitat Management Workshop*, Western Division, American Fisheries Society and Cooperative Fisheries Unit, Humboldt State University, Arcata, Calif., pp. 222–245.

Sedell, J.R., J.E. Yuska, and R.W. Speaker. 1984. Habitats and salmonid distribution in pristine, sediment-rich river valley systems: S. Fork Hoh and Queets River, Olympic National Park. *in Fish and Wildlife Relationships in Old-Growth Forests*, W.R. Meehan, T.R. Merrell, and T.A. Hanley, Jr. (Eds.), American Institute of Fisheries Research Biology, pp. 33–46.

Sedell, J.R. and J.L. Froggatt. 1984. Importance of streamside forests to large rivers: The isolation of the Willamette River, Oregon, U.S.A., from its floodplain by snagging and streamside forest removal. *Verhandl. Internatl. Verein. Theor. Angew. Limnol.* (International Association of Theoretical and Applied Limnology) 22:1828–1834.

Sedell, J.R. and K.J. Luchessa. 1982. Using the historical record as an aid to salmonid habitat enhancement. *in Acquisition and Utilization of Aquatic Habitat Inventory Information*, N.B. Armantrout (Ed.), Western Oregon Division, American Fisheries Society, pp. 210–223.

Sedell, J.R., P.A. Bisson, F.J. Swanson, and J.V. Gregory. 1988. What we know about large trees that fall into streams and rivers. *in From the Forest to the Sea, A Story of Fallen Trees*, C. Maser, R.F. Tarrant, J.M. Trappe, and J.F. Franklin (Tech. Eds.), USDA Forest Service General Technical Report GTR-PNW-229, Pacific Northwest Research Station, Portland, Ore., pp. 47–81.

Sedell, J.R., P.A. Bisson, J.A. June, and R.W. Speaker. 1982. Ecology and habitat requirements of fish populations in South Fork Hoh River, Olympic National Park. *in Ecological Research in National Parks of the Pacific Northwest*, E.E. Starkey, J.F. Franklin, and J.W. Matthews (Eds.), Forestry Research Laboratory, Oregon State University, Corvallis, pp. 47–63.

Sedell, J.R. and W.S. Duval. 1985. Water transportation and storage of logs. *in Influence of Forest and Rangeland Management on Anadromous Fish Habitat in Western North America*, W.R. Meehan (Ed.), USDA Forest Service General Technical Report PNW-186, Pacific Northwest Forestry and Range Experiment Station, Portland, Ore., pp. 1–68.

Sedell, J.R., G.H. Reeves, R.F. Hauer, J.A. Stanford, and C.P. Hawkins. 1990. Role of refugia in recovery from disturbances: Modern fragmented and disconnected river systems. *Environ. Manage.* 14:711–724.

Sedell, J.R. and R.L. Beschta. 1991. Bringing back the "bio" in bioengineering. in *Fisheries Bioengineering Symposium,* J. Colt and R.J. White (Eds.), American Fisheries Society, Bethesda, Md., pp. 160–175.

Statzner, B. and B. Higler. 1985. Questions and comments on the river continuum concept. *Can. J. Fish. Aquat. Sci.* 42:1038–1044.

Stebbins, R.C. 1954. *Amphibians and Reptiles of Western North America,* McGraw-Hill, New York, 528 pp.

Steedman, R.J. 1983. Life History and Feeding Role of the Xylophagous Aquatic Beetle, *Lara avara* LeConte (Dryopoidea:Elmidae), M.S. thesis, Oregon State University, Corvallis, 106 pp.

Steedman, R.J. and N.H. Anderson. 1985. Life history and ecological role of the xylophagous aquatic beetle, *Lara avara* LeConte (Dryopoidea:Elmidae). *Freshwater Biol.* 15:535–546.

Stein, R.A., P.E. Reimers, and J.D. Hall. 1972. Social interaction between juvenile coho (*Oncorhynchus kisutch*) and fall Chinook salmon (*O. tschawytscha*) in Sixes River, Oregon. *J. Fish. Res. Board Can.* 29:1737–1748.

Stembridge, J.E., Jr. 1979. Beach protection properties of accumulated driftwood. in *Proceedings of the Specialty Conference on Coastal Structures 79,* ASCE, Alexandria, Va., pp. 1052–1068.

Suyehiro Y. 1952. *Textbook of Ichthyology* [in Japanese], Iwanami Shoten, Tokyo, 332 pp.

Suzuki, Z. 1992. General Description on Tuna Biology Related to Fishing Activities on Floating Objects by Japanese Purse Seine Boats in the Western and Central Pacific, paper given at the Inter-American Tropical Tuna Commission, La Jolla, Calif., 9 pp.

Swanson, F.J. and D.N. Swanston. 1977. Complex mass-movement terrains in the western Cascade Range, Oregon. *Geol. Soc. Am. Rev. Eng. Geol.* 3:113–124.

Swanson, F.J. and G. Lienkaemper. 1978. Physical Consequences of Large Organic Debris in Pacific Northwest Streams, USDA Forest Service General Technical Report PNW-69, Pacific Northwest Forestry and Range Experiment Station, Portland, Ore., 12 pp.

Swanson, F.J. and G. Lienkaemper. 1982. Interactions among fluvial processes, forest vegetation, and aquatic ecosystems, South Fork Hoh River, Olympic National Park. in *Ecological Research in National Parks of the Pacific Northwest*, E.E. Starkey, J.F. Franklin, and J.W. Matthews (Eds.), Forestry Research Laboratory, Oregon State University, Corvallis, pp. 30–34.

Swanson, F.J., G. Lienkaemper, and J.R. Sedell. 1976. History, Physical Effects, and Management Implications of Large Organic Debris in Western Oregon Streams, USDA Forest Service General Technical Report PNW-56, Pacific Northwest Forestry and Range Experiment Station, Portland, Ore., 15 pp.

Swanson, F.J., M.D. Bryant, G.W. Lienkaemper, and J.R. Sedell. 1984. Organic Debris in Small Streams, Prince of Wales Island, Southeast Alaska, USDA Forest Service General Technical Report PNW-166, Pacific Northwest Forestry and Range Experiment Station, Portland, Ore., 12 pp.

Swanson, F.J., R.L. Graham, and G.E. Grant. 1985. Some effects of slope movements on river channels. in Proceedings International Symposium of Erosion, Debris Flow and Disaster Prevention, Tsukuba, Japan, 4:273–278.

Swanston, D.N. and F.J. Swanson. 1976. Timber harvesting, mass erosion, and steepland forest geomorphology in the Pacific Northwest. in *Geomorphology and Engineering*, D.R. Coates (Ed.), Dowden, Hutchinson, and Ross, Stroudsburg, Pa., pp. 199–221.

Terich, Thomas and Scott Milne. 1977. The Effects of Wood Debris and Drift Logs on Estuarine Beaches of Northern Puget Sound, Department of Geography and Regional Planning, Unpublished Project Report OWRT Agreement Number 143-34-10E-3996-5003, Western Washington University, Bellingham, 6 pp.

The Associated Press. 1989. Volunteers haul trash from beaches. *The Corvallis Gazette-Times* (Corvallis, Ore.), March 26.

Thomas, Duncan W. 1983. Changes in Columbia River Estuary Habitat Types Over the Past Century, Report of the Columbia River Estuary Data Development Program, 51 pp.

Tipper, R.C. 1968. Ecological Aspects of Two Wood-Boring Molluscs from the Continental Terrace off Oregon, Ph.D. thesis, Oregon State University, Corvallis, 137 pp.

Toews, D.A.A. and M.K. Moore. 1982. The Effects of Streamside Logging on Large Organic Debris in Carnation Creek, Land Management Paper 11, Ministry of Forests, Vancouver, B.C., 29 pp.

Toufexis, A. 1988. The dirty seas. *Time* 132:44–50.

Towle, J.C. 1974. Woodland in the Willamette Valley: An Historical Geography, Ph.D. dissertation, University of Oregon, Eugene.

Trappe, J.M. and C. Maser. 1977. Ectomycorrhizal fungi: Interactions of mushrooms and truffles with beasts and trees. in *Mushrooms and Man: An Interdisciplinary Approach to Mycology*, T. Walters (Ed.), Linn Benton Community College, Albany, Ore., pp. 165–178.

Triska, F.J. and K. Cromack, Jr. 1980. The role of wood debris in forests and streams. in *Forests: Fresh Perspectives from Ecosystem Analysis*, Proceedings of the 40th Annual Biology Colloquium, R.H. Waring (Ed.), Oregon State University Press, Corvallis, pp. 171–190.

Triska, F.J., J.R. Sedell, and S.V. Gregory. 1982. Coniferous forest streams. in *Analysis of Coniferous Forest Ecosystems in the Western United States*, R.L. Edmonds (Ed.), US/IBP Synth. Series, Hutchinson Ross, Stroudsburg, Pa., pp. 292–332.

Tschaplinski, P.J. and G.F. Hartman. 1983. Winter distribution of juvenile coho salmon (*Oncorhynchus kisutch*) before and after logging in Carnation Creek, British Columbia, and some implications for overwinter survival. *Can. J. Fish. Aquat. Sci.* 40:452–461.

Turner, R.D. 1956. Notes on *Xylophaga washingtona* Bartsch and on the genus. *Nautilus* 70:10–12.

Turner, R.D. 1971. Australian shipworms. *Austral. Nat. Hist.* 17:139–145.

Turner, R.D. 1977. Wood, mollusks, and deep-sea food chains. *Bull. Am. Malacol. Union* 1976:13–19.

Turner, R.D. 1981. "Wood islands" and "thermal vents" as centers for diverse communities in the deep sea. *Soviet J. Mar. Biol.* 7:3–29 (translation of *Biol. Morya* 7:3–10, 1981).

Uda, M. 1933. Types of skipjack schools and their fishing qualities [in Japanese]. *Bull. Jap. Soc. Sci. Fish.* 2:107–111 (English translation in Van Campen, W.G. 1952. Five Japanese papers on skipjack. *U.S. Fish Wild. Serv. Spec. Sci. Rep. Fish.* 83:68–73).

Vannote, R.L., G.W. Minshall, K.W. Cummins, et al. 1980. The river continuum concept. *Can. J. Fish. Aquat. Sci.* 37:130–137.

Vinogradov, M.E. and E.A. Shushkina. 1985. Some aspects of the study of ecosystems of the epipelagial of the ocean. in *Biological Basis of the Commercial Exploitation of the Open Areas of the Ocean*, M.E. Vinigradov and M.V. Flint (Eds.), Academy of Sciences of the U.S.S.R., Commission on Problems of the World Ocean, Nauka, Moscow, pp. 8–20.

Wallace, J.B. and A.C. Benke. 1984. Quantification of wood habitat in subtropical coastal plain streams. *Can. J. Fish. Aquat. Sci.* 41:1643–1652.

Waring, R.H. 1987. Characteristics of trees predisposed to die. *BioScience* 37:569–574.

Weibe, W.J. and J. Liston. 1972. Studies of the aerobic, nonexacting, heterotrophic bacteria of the benthos. in *The Columbia River Estuary and Adjacent Waters Bioenvironmental Studies,* A.T. Pruter and D.L. Alverson (Eds.), University of Washington Press, Seattle, pp. 281–312.

Weidemann, A.M., L.R.J. Dennis, and F.H. Smith. 1969. *Plants of the Oregon Coastal Dunes,* Oregon State University Book Stores, Corvallis, 117 pp.

Weisskopf, M. 1988. Plastic reaps a grim harvest in the oceans of the world. *Smithsonian* March:59–66.

Welcomme, R.L. 1985. *River Fisheries,* FAO Tech. Paper 262:1–330.

Whitton, B.A. 1984. *Ecology of European Rivers,* Blackwell Scientific, Oxford, England, 520 pp.

Wiggens, G.B. 1977. *Larvae of the North American Caddisfly Genera (Trichoptera),* University of Toronto Press, Toronto.

Xavier, Ariz, Delgado Alicia, Fonteneau Alain, Gonzales Costas, Fernando Pilar, and Pallares Pilar. 1992. Logs and Tunas in the Eastern Tropical Atlantic, A Review of Present Knowledge and Uncertainties, paper given at the Inter-American Tropical Tuna Commission, La Jolla, Calif., 23 pp.

Yabe, H. and T. Mori. 1950. An observation on a school of skipjack and Kimeji accompanying a drift log [in Japanese with English summary]. *J. Jap. Soc. Sci. Fish.* 16:35–39.

Younger, L.K. and K. Hodge. 1992. *1991 International Coastal Cleanup Results,* R. Bierce and K. O'Hara (Eds.), Center for Marine Conservation, Washington, D.C., 470 pp.

Zimmerman, R.C., J.C. Goodlet, and G.H. Comer. 1967. The influence of vegetation on channel form of small streams. in *Symposium on River Morphology,* Internatl. Assoc. Sci. Hydrol. Publ. 75, Gentbrugge, Belgium, pp. 225–275.

INDEX